ULTIMATE
BRAINBUSTER

FILM&
TV QUIZ
BOOK

ULTIMATE BRAINBUSTER | FILM & TV QUIZ BOOK

JON BARRENECHEA | NIKKI BAYLEY

ILLUSTRATIONS by VICTORIA WAINWRIGHT

Published by Igloo Books Limited
Henson Way
Telford Way Industrial Estate
Kettering
Northants
NN16 8PX

info@igloo-books.com

This edition published 2005

ISBN 1-84561-198-5

Project management: Kandour Ltd
Editorial and design management: boxedBRAND
Author: Jon Barrenechea with Nikki Bayley
Illustrations: Victoria Wainwright
Cover design: Peter Gates
Production: Jane Utting

CONTENTS

ROUND 1 | FILM DIVAS
8 - 27

ROUND 2 | ACTION FILMS
28 - 45

ROUND 3 | THRILLERS
46 - 63

ROUND 4 | HORROR
64 - 83

ROUND 5 | CHILDREN'S FILMS
84 -103

ROUND 6 | SOAP OPERAS
104 -121

ROUND 7 | COPS & ROBBERS
122 -141

ROUND 8 | ROMANCE
142 -161

ROUND 9 | MUSICALS
162 - 179

ROUND 10 | ALTERNATIVE CINEMA
180 - 199

ROUND 11 | SCIENCE FICTION
200 - 219

ROUND 12 | CULT FILM & TV
220 - 237

ROUND 13 | COMEDIES
238 - 257

ROUND 14 | DIRECTORS
258 - 277

ROUND 15 | AWARD WINNERS
278 - 295

ROUND 16 | RANDOM
296 - 317

THE BRAINBUSTERS
318 - 320

ULTIMATE
BRAINBUSTER

ROUND 1
FILM

DIVAS

QUESTIONS

1. Which actress has been linked romantically to Jonny Lee Miller, Billy Bob Thornton and Brad Pitt?
2. Greta Garbo never actually said "I want to be alone", what did she say?
3. Which singing diva flopped at the box office with Glitter?
4. What was the name of Whitney Houston's character in The Bodyguard?
5. Which drama school did Joan Collins attend?
6. What was the name of Laurence Olivier's second wife?
7. Which actress links Don't Look Now, Fahrenheit 451 and Billy Liar?
8. Which actress links a giant ape and a tragic country and western singer?
9. To which playwright was Marilyn Monroe briefly married?
10. Before marrying Coldplay's Chris Martin, Gwyneth Paltrow had been romantically-linked to which Se7en co-star?

ANSWERS

1. Angelina Jolie
2. I want to be left alone.
3. Mariah Carey
4. Rachel 'Rach' Marron
5. RADA
6. Joan Plowright
7. Julie Christie
8. Jessica Lange
9. Arthur Miller
10. Brad Pitt

ROUND 1
FILM DIVAS 2

1. Who took the starring role in a Patsy Cline biopic?
2. Who co-starred with Cary Grant in North by Northwest?
3. Zsa Zsa Gabor is most identified with what item of jewelry?
4. Which playwright was briefly married to Marilyn Monroe?
5. Which rap mogul was Jennifer Lopez associated with?
6. What part of her body did Goldie Hawn stipulate MUST feature naked in films?
7. Who played Lebowski's daughter in The Big Lebowski?
8. Name the 1920s starlet famed for her severe black bob and her veracious sexual appetite.
9. Nicole Kidman starred in BMX Bandits, true or false?
10. What is the name of Gwyneth Paltrow's husband's band's third album?

1. Jessica Lange
2. Eve Marie Saint
3. Diamond necklaces
4. Arthur Miller
5. Sean Coombes aka Puff Daddy aka P. Diddy
6. Her bottom
7. Julianne Moore
8. Louise Brooks
9. True
10. X&Y

QUESTIONS

1. In what film does Katharine Hepburn fall in love in Italy while on holiday?
2. Jane Fonda married which media mogul after retiring in 1991?
3. In which films did Liza Minnelli star opposite Dudley Moore?
4. Liza Minnelli is the daughter of Judy Garland and which director?
5. Of which Principality did Grace Kelly become Princess?
6. What profession did Elizabeth Taylor's character have in the film Butterfield 8?
7. What was the name of Audrey Hepburn's character in Breakfast at Tiffany's?
8. In which 1946 film does Rita Hayworth seductively take her gloves off?
9. To which filmmaker was Rita Hayworth married to in the 1940s?
10. Marilyn Monroe sang 'Happy Birthday' to which American President?

ANSWERS

1. Summertime
2. Ted Turner
3. Arthur 1 & 2
4. Vincent Minnelli
5. Monaco
6. Call girl
7. Holly Golightly
8. Gilda
9. Orson Welles
10. John F. Kennedy

ROUND 1
FILM DIVAS 4

QUESTIONS

1. In which film did Eve Marie Saint appear with Marlon Brando?
2. With which younger actor is Demi Moore romantically-linked?
3. Cabaret starred Liza Minelli. But what is the name of the book upon which it was based?
4. What was the first Tim Burton film to feature Winona Ryder?
5. In Batman Returns, Michelle Pfeiffer modifies a neon sign to say "Hell Here". What did it say originally?
6. Which English town has a street named for Kate Winslett?
7. Vanessa Redgrave starred in which cult 1966 movie with David Hemmings?
8. Who is the actress daughter of the late Sir John Mills?
9. … and her son was the lead singer of which 1990s pop group?
10. Which actress plays opposite Gregory Peck in the Western Duel in the Sun?

ANSWERS

1. On The Waterfront
2. Ashton Kutcher
3. I Am A Camera
4. Beetlejuice
5. Hello There
6. Reading
7. Blow Up
8. Hayley Mills
9. Kula Shaker
10. Jennifer Jones

1. In what year did Katherine Hepburn die?
2. In which state was Annette Bening born: Kansas, Florida or Delaware?
3. Was Judi Dench born in York or New York?
4. Which Bond movie featured Halle Berry?
5. On which Spanish island to Catherine Zeta Jones and Michael Douglas have a retreat?
6. What is the name of Eric Roberts' more famous sister?
7. Which actress was one of the co-founders of United Artists?
8. Who played the lead role in Pretty In Pink?
9. Aviator star Gwen Stefani has her own clothin label. What is called?
10. The Jane Fonda-starring movie, Barbarella, included a character whose name was borrowed by an internationally -successful pop group. Which pop group?

10.	Duran Duran	5.	Mallorca
9.	L.A.M.B.	4.	Die Another Day
8.	Molly Ringwald	3.	York
7.	Mary Pickford	2.	Kansas
6.	Julia Roberts	1.	2003

ROUND 1
FILM DIVAS 6

1. Which Spanish actress replaced Nicole Kidman as Tom Cruise's love interest?
2. Which child star of E.T went on to star in the film Charlie's Angels?
3. Which famous actress was only three years old when she appeared in the film The Red-Haired Alibi?
4. What university did Jodie Foster attend?
5. Jennifer Jason Leigh is the daughter of which actor?
6. What is Bridget Fonda's relation to Jane Fonda?
7. Who was Ginger Rogers' most famous dancing partner?
8. Which former model rose to fame opposite Jim Carrey in The Mask?
9. In which film is Ann-Margret the love interest of two senior citizens?
10. Kate Winslet is married to which acclaimed director?

1. Penelope Cruz
2. Drew Barrymore
3. Shirley Temple
4. Yale
5. Vic Morrow
6. Niece
7. Fred Astaire
8. Cameron Diaz
9. Grumpy Old Men
10. Sam Mendes

1. To what actor is Sarah Jessica Parker married?
2. In which country was Kim Cattrall born?
3. Heather Locklear is married to the guitar player from which rock band?
4. Pamela Anderson was married to the drummer from which rock band?
5. What crime was Winona Ryder arrested for in 2001?
6. Which rock star is Liv Tyler the daughter of?
7. Alicia Silverstone starred in which modern adaptation of a Jane Austen book?
8. Courteney Cox appeared in which Bruce Springsteen music video?
9. In which year did Jennifer Lopez launch her music career?
10. In which film did Jennifer Lopez play a murdered Mexican pop star?

1. Matthew Broderick
2. England
3. Bon Jovi
4. Motley Crue
5. Shoplifting
6. Steve Tyler from Aerosmith
7. Clueless
8. Dancing in the Dark
9. 1998
10. Selena

ROUND 1
FILM DIVAS 8

1. Which silent film star was known as 'America's Sweetheart'?
2. Which teenage actress was the lead in the film Mean Girls?
3. Which teenage actress/pop star was the lead in the Lizzie Maguire Movie?
4. In Sunset Boulevard, which role does Gloria Swanson play?
5. What was Marilyn Monroe's last film?
6. In which film does Audrey Hepburn play a princess on the run in Rome?
7. Elizabeth Taylor had a turbulent marriage to which British actor?
8. In how many films did Doris Day star with Rock Hudson?
9. Tippi Hedren is the mother of which actress?
10. Carrie Fisher is the daughter of which actress?

6.	Roman Holiday	1.	Mary Pickford	
7.	Richard Burton	2.	Lindsay Lohan	
8.	Three	3.	Hilary Duff	
9.	Melanie Griffith	4.	Norma Desmond	
10.	Debbie Reynolds	5.	The Misfits	

FILM DIVAS 9

1. Frances Farmer's life was made into a film, Frances. Which actress played the title role?
2. Who plays Mallory in Natural Born Killers?
3. For which other film star did Eddie Fisher leave Debbie Reynolds?
4. Jane Fonda travelled to which country in the 1970s, earning her the nickname Hanoi Jane?
5. Susan Sarandon starred in which 1975 transvestite musical?
6. Geena Davis won an Oscar for her performance in which film?
7. The Long Kiss Goodnight was a Geena Davis film directed by her ex-husband. What's his name?
8. How is Hilary Swank related to Rob Lowe?
9. Jamie Lee Curtis is the daughter of which actress?
10. Heather Graham starred as 'Rollergirl' in which film?

1. Jessica Lange
2. Juliette Lewis
3. Elizabeth Taylor
4. Vietnam
5. Rocky Horror Picture Show
6. Thelma & Louise
7. Renny Harlin
8. Sister-in-law
9. Janet Leigh
10. Boogie Nights

18   FILM & TV QUIZ BOOK

ROUND 1
FILM DIVAS 10

1. What was Cybil Shepherd's film debut?
2. Which David Lynch film features Laura Dern as the lead opposite Nicolas Cage?
3. Where was Naomi Watts born?
4. Who plays the stalker in Fatal Attraction?
5. Lauren Bacall married which movie star when she was 20 years old?
6. Jane Russell came to fame as the busty star of which Howard Hughes film?
7. In which film did Jane Russell and Marilyn Monroe star together?
8. What country was Ingrid Bergman from?
9. Isabella Rossellini is the daughter of which star?
10. Who does Burt Lancaster roll in the sand with in From Here to Eternity?

1. Last Picture Show
2. Wild at Heart
3. England
4. Glenn Close
5. Humphrey Bogart
6. The Outlaw
7. Gentlemen Prefer Blondes
8. Sweden
9. Ingrid Bergman
10. Deborah Kerr

QUESTIONS

1. Which silent film star's name is an anagram for Arab Death?
2. Which actress said 'When I'm good, I'm very good, but when I'm bad, I'm better'?
3. Vivien Leigh was married to which British Academy Award-winning actor/director?
4. Which baseball legend did Marilyn Monroe marry?
5. In which controversial film does Shelly Winters star opposite James Mason?
6. What film featured Whitney Houston as a pop star whose life is threatened?
7. Ralph Fiennes starred opposite which actress in the film Maid in Manhattan?
8. Which commercially and critically disastrous film did Ben Affleck and Jennifer Lopez collaborate on?
9. In which film does Salma Hayek dance on stage in front of George Clooney and Quentin Tarantino?
10. Linda Fiorentino plays a femme fatale in which 1994 film?

ANSWERS

1. Theda Bara
2. Mae West
3. Lawrence Olivier
4. Joe DiMaggio
5. Lolita
6. The Bodyguard
7. Jennifer Lopez
8. Gigli
9. From Dusk till Dawn
10. The Last Seduction

ROUND 1
FILM DIVAS 12

QUESTIONS

1. Which silent film actress was mostly associated with D.W. Griffith and starred in Birth of a Nation and Intolerance?
2. The autobiography Lulu in Hollywood was written by which silent film star?
3. Which actress played the object of Jimmy Stewart's obsession in Vertigo?
4. Maureen O'Hara starred opposite John Wayne in which John Ford Irish film?
5. Who played Jane in all the original Tarzan films from the 1930s?
6. In what film does Meg Ryan play an alcoholic?
7. What films stars Goldie Hawn and Susan Sarandon as former groupies?
8. Which British actress plays Jean Brodie in The Prime of Miss Jean Brodie?
9. For which film did Mira Sorvino win an Academy Award?
10. Sissy Spacek made her debut in which 1973 film opposite Martin Sheen?

ANSWERS

1. Lillian Gish
2. Louise Brooks
3. Kim Novak
4. The Quiet Man
5. Maureen Sullivan
6. When a Man Loves a Woman
7. The Banger Sisters
8. Maggie Smith
9. Mighty Aphrodite
10. Badlands

1. Which silent film star did Mary Pickford marry?
2. Norma Shearer was a protégé of which powerful film mogul?
3. Which actress played Cleopatra on screen in 1934, before Elizabeth Taylor?
4. What was the name of Joan Crawford's first and last Western?
5. Who stars opposite William Powell in My Man Godfrey?
6. Who plays Mary Jane in the Spiderman films?
7. Who stars as Mrs Robinson in The Graduate?
8. Which French actress shocked the world with the film ...And God Created Woman?
9. Which child actress starred as Elliot's little sister in E.T.?
10. For which film did Kim Basinger win an Academy Award?

10. L.A. Confidential
9. Drew Barrymore
8. Bridgette Bardot
7. Anne Bancroft
6. Kirsten Dunst

5. Carole Lombard
4. Johnny Guitar
3. Claudette Colbert
2. Irving Thalberg
1. Douglas Fairbanks Snr.

ROUND 1
FILM DIVAS 14

QUESTIONS

1. Which actress plays Eve Harrington in All About Eve?
2. Who was the first actress to be called the 'It' girl?
3. Ellen Burstyn starred in which classic horror film?
4. What bus-centred film made Sandra Bullock a star?
5. Who plays Cruella DeVille in the live-action version of 101 Dalmatians?
6. For which film did Jennifer Connelly win an Academy Award in 2000?
7. Which teenage actress did Jamie Lee Curtis star opposite in Freaky Friday?
8. Who is teenage star Hilary Duff's showbiz rival?
9. Whose autobiography Postcards from the Edge was turned into a Meryl Streep film?
10. Jennifer Aniston married which Hollywood sex symbol in 2000?

ANSWERS

1. Anne Baxter
2. Clara Bow
3. The Exorcist
4. Speed
5. Glenn Close
6. A Beautiful Mind
7. Lindsay Lohan
8. Lindsay Lohan
9. Carrie Fisher
10. Brad Pitt

QUESTIONS

1. What film starred Joan Crawford and Greta Garbo?
2. For which film did Bette Midler receive an Academy Award nomination?
3. What film brought Goldie Hawn, Diane Keaton and Bette Midler together?
4. What film united Jane Fonda, Lily Tomlin and Dolly Parton?
5. In which film does Lana Turner play an actress and Kirk Douglas a producer?
6. The Postman Always Rings Twice was a Lana Turner film remade with whom in the Turner part?
7. What was the name of the studios founded by Lucille Ball and husband Desi Arnaz?
8. What was Romy Schneider's cause of death?
9. Which French actress starred in films such as Umbrellas of Cherbourg, The Last Metro and Indochine?
10. For which film did Juliette Binoche win an Oscar?

ANSWERS

1. Grand Hotel
2. The Rose
3. First Wives Club
4. 9 to 5
5. The Bad and the Beautiful
6. Jessica Lange
7. Desilu
8. Cardiac arrest
9. Catherine Deneuve
10. The English Patient

ROUND 1
FILM DIVAS 16

1. Kate Beckinsale is the daughter of which late British TV actor?
2. In what film do Neve Campbell, Denise Richards and Matt Dillon have a threesome?
3. Who plays Juliet opposite Leo DiCaprio in Baz Luhrman's film?
4. Charlize Theron won an Oscar for her performance as which serial killer?
5. In which film does Renee Zellweger tell Tom Cruise, 'You had me at hello'?
6. Which Latin actress stars in the TV show Dark Angel?
7. What was R&B artist Aaliyah's last film?
8. Emma Thompson was married to which actor/director?
9. Who stars as Laura in the film of the same name?
10. Who plays Elle Woods in Legally Blonde?

1. Richard Beckinsale
2. Wild Things
3. Claire Danes
4. Aileen Wuornos
5. Jerry Maguire
6. Jessica Alba
7. The Queen of the Damned
8. Kenneth Branagh
9. Gene Tierney
10. Reese Witherspoon

QUESTIONS

1. Who played Lolita's mom in the 1960 Stanley Kubrick film?
2. What was Natalie Wood's character called in Rebel Without A Cause?
3. Who was the damsel in distress in King Kong (1933)?
4. Which actress played Joseph Cotton's niece in Hitchcock's Shadow of a Doubt?
5. Catherine Zeta Jones started out in which British TV series?
6. What is the name of Julia Roberts' older brother, also an actor?
7. Who played the title role in the film Julia?
8. Which red-headed actress starred in many of John Hughes' teenage comedies of the 1980s?
9. Which South American musical's star carried a lot of fruit on her head?
10. Which actress took it all off for the film Striptease?

ANSWERS

1. Shelly Winters
2. Judy
3. Fay Wray
4. Teresa Young
5. The Darling Buds of May
6. Eric Roberts
7. Vanessa Redgrave
8. Molly Ringwald
9. Carmen Miranda
10. Demi Moore

ROUND 1
FILM DIVAS 18

QUESTIONS

1. Who plays Robert Redford's girlfriend in Butch Cassidy and the Sundance Kid?
2. Who was Gena Rowlands' husband and frequent director?
3. Who starred opposite Cary Grant in His Girl Friday?
4. What was the only film Meg Ryan and husband Dennis Quaid made together?
5. In which high school film does Winona Ryder team up with Christian Slater?
6. Eva Marie Saint was leading lady to which actor in North by Northwest?
7. Kathleen Turner played the femme fatale opposite William Hurt in which 1981 film?
8. Angelica Huston is the daughter of which famous director?
9. Which member of her family did Angelina Jolie bring to the Oscars on the night she won an Award in 1999?
10. Which actress plays opposite Gregory Peck in the Western Duel in the Sun?

ANSWERS

1. Katherine Ross
2. John Cassavetes
3. Rosalind Russell
4. D.O.A.
5. Heathers
6. Cary Grant
7. Body Heat
8. John Huston
9. Her brother
10. Jennifer Jones

ULTIMATE
BRAINBUSTER

ROUND 2
ACTION FILMS

1. What was the first James Bond film?
2. What is the name of the high-rise building where John McClane is trapped with the terrorists in Die Hard?
3. What famous restaurant chain has one of its locations destroyed at the end of Con Air?
4. In how many Bond films has Roger Moore played the title role?
5. Samuel Jackson stars as Zeus, McClane's reluctant sidekick, in Die Hard with a Vengeance. In which film had Bruce Willis and Samuel Jackson previously both starred?
6. Which actor plays James Bond in On Her Majesty's Secret Service?
7. Con Air and The Rock share a producer in common, a man known for his over- the-top action films. What's his name?
8. At what speed will the bus explode in the film Speed?
9. The Terminator comes from what year in the future in the first film?
10. What was the name of the first Rambo film?

1. Dr No
2. Nakatomi Plaza
3. Hard Rock Cafe
4. Seven
5. Pulp Fiction
6. George Lazenby
7. Jerry Bruckheimer
8. 50 mph
9. 2029
10. First Blood

ROUND 2
ACTION FILMS 2

QUESTIONS

1. To which Western classic does Hans Gruber (erroneously) refer to at the end of Die Hard?
2. In which film does Tom Cruise move to Japan and become a Samurai?
3. In which location does the first Die Hard sequel Die Harder take place?
4. The Rock refers to which now decommissioned penitentiary facility in California?
5. What was Bruce Lee's first film?
6. Nicolas Cage took the role of Dr Stanley Godspeed in The Rock directly after winning the Oscar for what film?
7. What is the name of the author on which the James Bond films are based?
8. What country does John Rambo travel to in Rambo III?
9. Who plays Indiana Jones' father in the third film of the series?
10. What goes missing in The Sum of All Fears?

ANSWERS

1. High Noon
2. The Last Samurai
3. An Airport
4. Alcatraz
5. Fists of Fury
6. Ian Fleming
7. Afghanistan
8. Tom Selleck
9. Sean Connery
10. A nuclear weapon

QUESTIONS

1. Which actor replaced Keanu Reeves in the sequel Speed 2: Cruise Control?
2. Which actress stars as Tom Cruise's love interest in Mission Impossible: 2?
3. What model is the robot sent to kill Connor in Terminator 2: Rise of The Machines?
4. In Under Siege, what is Steven Seagal's position on the aircraft carrier that is hijacked?
5. In True Lies, what does Arnold Schwarzenegger tell his wife he does for a living?
6. In the Rambo films, Sylvester Stallone plays John Rambo, a veteran of what war?
7. What Royal Academy of Dramatic Arts-trained British actor played terrorist Hans Gruber in Die Hard?
8. Casino Royale is a James Bond spoof starring which actor as James Bond's nephew?
9. What was Pierce Brosnan's first James Bond film?
10. Who plays Jack Ryan in Hunt for Red October?

ANSWERS

1. Jason Patric
2. Thandie Newton
3. T-1000
4. Cook
5. Computer Salesman
6. Vietnam
7. Alan Rickman
8. Woody Allen
9. Goldeneye
10. Alec Baldwin

ROUND 2
ACTION FILMS 4

QUESTIONS

1. What Jackie Chan film is set in the Bronx?
2. Chuck Norris made a series of Rambo-style films in the 1980s – what was their name?
3. What actress plays Lara Croft?
4. In which film do Gregory Peck and Anthony Quinn plot to blow up a Nazi base?
5. In which film does Lee Marvin lead twelve criminals on a WW2 mission?
6. Assault on Precinct 13 was remade with which actor in the lead?
7. What character does Kurt Russell play in Escape from New York?
8. In what film do Geena Davis and Susan Sarandon go on a crime spree?
9. What Bond film's plot has him walking into a trap in Istanbul set by SPECTRE?
10. Which Hong Kong director helmed the sequel Mission Impossible 2?

ANSWERS

1. Rumble in the Bronx
2. Missing in Action
3. Angelina Jolie
4. Guns of Navarone
5. Dirty Dozen
6. Ethan Hawke
7. Snake Plissen
8. Thelma & Louise
9. From Russia With Love
10. John Woo

QUESTIONS

1. To which country does Jack Ryan travel to in Clear and Present Danger?
2. Who plays the President of the United States in the film Air Force One?
3. Who composed the theme song for the film Mission: Impossible and which artist reworked it in 1996?
4. What series of revenge films did Charles Bronson make in the 1970s and 1980s?
5. In which famous tunnel does the spectacular finale of Mission Impossible take place?
6. Who is the star of the film XXX?
7. What is the name of the actor who plays 'Jaws' in the Roger Moore Bond films?
8. What precious archaeological object are the Nazis after in Indiana Jones: The Last Crusade?
9. What British Academy Award-winning actor plays the villain in Die Hard's third outing, Die Hard with a Vengeance?
10. Who was originally set to play Indiana Jones before Harrison Ford?

ANSWERS

1. Colombia
2. Harrison Ford
3. Lalo Schifrin and Bono/ The Edge
4. Death Wish
5. Channel Tunnel
6. Vin Diesel
7. Richard Kiel
8. The Holy Grail
9. Jeremy Irons
10. Tom Selleck

ROUND 2
ACTION FILMS 6

1. Which actor was the most famous swashbuckling hero of the 1930s and 1940s?
2. What film paired Sean Connery and Michael Caine for the first time?
3. What modern pirate film stars Geena Davis?
4. King Kong was remade with Jeff Bridges and Jessica Lange in what decade?
5. In what desert does the plane crash in The Flight of the Phoenix?
6. What actor plays the title role in Lawrence of Arabia?
7. In which film does Daniel Day Lewis play Hawkeye?
8. For which film did Mel Gibson win Best Picture Oscar?
9. In which film does Sylvester Stallone play a mountain climber?
10. Which actor plays the villain in River Wild?

1. Errol Flynn
2. The Man Who Would Be King
3. Cutthroat Island
4. 1970s (1976)
5. Sahara
6. Peter O'Toole
7. Last of the Mohicans
8. Braveheart
9. Cliffhanger
10. Kevin Bacon

QUESTIONS

1. Joe Don Baker stars as a real-life Sheriff in which series of vigilante films from the 1970s?
2. Who starred in the 2004 remake on Walking Tall?
3. What is the name of the actress who plays Jack Bauer's daughter on TV series 24?
4. In which film does Bruce Willis play a hired assassin?
5. What was Steven Seagal's previous occupation?
6. What sword-wielding character did Arnold Schwarzenegger play in two films?
7. What was the sequel to Mad Max?
8. What film stars Michael Douglas and Kathleen Turner searching for an Emerald in South America?
9. Who was the first actor to play Robin Hood on screen?
10. What actor played Tarzan on screen most often?

ANSWERS

10. Johnny Weissmuller
9. Douglas Fairbanks Sr
8. Romancing The Stone
7. Road Warrior
6. Conan

5. Bodyguard
4. The Jackal
3. Elisha Cuthbert
2. The Rock
1. Walking Tall

ROUND 2
ACTION FILMS 8

QUESTIONS

1. Which actor plays Jack Sparrow in Pirates of the Caribbean?
2. What is the name of the President of the US in the second season of 24?
3. In which year do the events depicted in the film Pearl Harbor take place?
4. In which film does Mel Gibson fight the English in the American Revolutionary war?
5. Three Kings is set during which war?
6. What Academy Award-winning actress was a Bond girl in Die Another Day?
7. In what city was Predator 2 set?
8. What films star Jackie Chan and Chris Tucker?
9. Where does Speed 2: Cruise Control take place?
10. Which car model was used in The Italian Job?

ANSWERS

1. Johnny Depp
2. Palmer
3. 1941
4. The Patriot
5. Gulf War of 1991
6. Halle Berry
7. Los Angeles
8. Rush Hour 1 & 2
9. On board a cruise
10. Mini Coopers

QUESTIONS

1. Which actress plays secret agent Sydney Bristow in Alias?
2. In which film does Denzel Washington save a little girl from kidnappers in Mexico?
3. What is the name of Uma Thurman's character in Kill Bill?
4. What John Le Carre film adaptation features Pierce Brosnan as a British spy?
5. What is the name of the ship commanded by Russell Crowe in Master & Commander?
6. In what film does Kiera Knightley play Guinevere?
7. Who plays Private Ryan in Saving Private Ryan?
8. In what film does Jude Law play a Soviet sniper in World War II?
9. During what war was Apocalypse Now set?
10. In what film do Robert De Niro and Christopher Walken play Russian roulette?

ANSWERS

1. Jennifer Garner
2. Man on Fire
3. The Bride
4. The Tailor of Panama
5. HMS Surprise
6. King Arthur
7. Matt Damon
8. Enemy At The Gates
9. Vietnam
10. Deer Hunter

ROUND 2
ACTION FILMS 10

QUESTIONS

1. What Brad Pitt film is an adaptation of Homer's The Iliad?
2. In which film does Tom Cruise learn to be a Samurai?
3. Which actor plays Sharpe in the TV series?
4. What 1996 film starring Helen Hunt depicts the devastating effect of tornadoes?
5. Who stars as the Michael Caine character in the 2003 remake of The Italian Job?
6. In which film does Denzel Washington battle terrorists in New York City?
7. Who plays Max in the TV series Dark Angel?
8. What actor plays the Man With No Name in the Spaghetti Western trilogy of Sergio Leone?
9. What is the name of the third film in the Mariachi trilogy of director Robert Rodriguez?
10. In what film does Pierce Brosnan escape an erupting volcano?

ANSWERS

1. Troy
2. The Last Samurai
3. Sean Bean
4. Twister
5. Mark Wahlberg
6. The Siege
7. Jessica Alba
8. Clint Eastwood
9. Once Upon a Time in Mexico
10. Dante's Peak

QUESTIONS

1. In which action film does Halle Berry show Hugh Jackman her breasts?
2. Who plays the voice of the dragon in Dragonheart?
3. In which country do the events depicted in Black Hawk Down take place?
4. What film based on a video game stars Jean-Claude Van Damme and Kylie Minogue?
5. In which Robin Hood film does Mary Elizabeth Mastrantonio play Marion?
6. Who starred with Nicolas Cage in the film Gone in 60 Seconds?
7. In which film does George Clooney play Lt. Col Thomas Devoe?
8. In which film does Steve McQueen perform multiple motorcycle stunts himself?
9. The Fast and the Furious is set around what underworld?
10. What actor plays Zorro in The Mask of Zorro?

ANSWERS

1. Swordfish
2. Sean Connery
3. Somalia
4. Street Fighter
5. Robin Hood: Prince of Thieves
6. Angelina Jolie
7. The Peacemaker
8. The Great Escape
9. Racing
10. Antonio Banderas

ROUND 2
ACTION FILMS 12

QUESTIONS

1. Tobey Maguire plays which superhero?
2. The director of Die Hard also directed Predator – what's his name?
3. In Escape from New York, what has Manhattan been turned into in the future?
4. What planet does Arnold Schwarzenegger travel to in Total Recall?
5. In which film does Harrison Ford play a Soviet submarine Captain?
6. What was Jean-Claude Van Damme's debut film?
7. What is the sequel to Escape from New York?
8. What film pairs Jackie Chan with Owen Wilson in the Old West?
9. In The Last Boy Scout, who does Bruce Willis team up with?
10. On what form of transportation does Under Siege 2 take place?

ANSWERS

1. Spider-Man
2. John McTiernan
3. A prison
4. Mars
5. K-19
6. Bloodsport
7. Escape from L.A.
8. Shanghai Noon
9. Damon Wayans
10. Train

QUESTIONS

1. What animals are at the centre of the film Deep Blue Sea?
2. What two films about a meteorite crashing into earth were released in 1998?
3. Who starred in A Knight's Tale as a jousting champion?
4. What was Pierce Bronson's last film as James Bond?
5. What Romeo & Juliet adaptation is a Jet Li martial arts film?
6. Which actor died while making the film The Crow?
7. In which films do Eddie Murphy and Nick Nolte pair up?
8. Mel Gibson and Robert Downey Jr are pilots for a secret airline in which film?
9. In which film does Jennifer Lopez battle a giant snake?
10. What number one single from Aerosmith was on the soundtrack to Armageddon?

ANSWERS

1. Sharks
2. Armageddon and Deep Impact
3. Heath Ledger
4. Die Another Day
5. Romeo Must Die
6. Brandon Lee
7. 48 Hrs and Another 48 Hrs
8. Air America
9. Anaconda
10. I Don't Want to Miss a Thing

QUESTIONS

1. What actor plays Clark Kent in the series Smallville?
2. What does CTU stand for on the series 24?
3. Mr. T starred in which TV series?
4. Mickey and Mallory are serial killers in what film?
5. What weapon does Bruce Willis use to rescue Ving Rhames with in Pulp Fiction?
6. Who plays Angelina Jolie's father in Tomb Raider
7. Who played Doc Holliday in the film Tombstone?
8. Who did Martin Sheen replace as the lead of the film Apocalypse Now?
9. Who plays 'The Bad' in The Good, The Bad and The Ugly?
10. Tom Cruise is a fighter pilot in which 1986 film?

ANSWERS

1. Tom Welling
2. Counter Terrorist Unit
3. The A-Team
4. Natural Born Killers
5. A sword
6. Jon Voight
7. Val Kilmer
8. Harvey Keitel
9. Lee Van Cleef
10. Top Gun

1. In Enemy of the State, what is Will Smith's profession?
2. What is the name of Uma Thurman's vehicle in Kill Bill 1?
3. Who directed The Wild Bunch?
4. Which actor played Ben Hur?
5. Which actor plays The Ugly in The Good, the Bad and the Ugly?
6. Who plays the Nameless Killer in Hero?
7. Mark Wahlberg and Chow-Yun Fat star in what film together?
8. The Last Man Standing, starring Bruce Willis, is a remake of which Spaghetti Western?
9. In which Western does John Wayne rescue Natalie Wood from Indians?
10. Who stars in Phone Booth?

1. Lawyer
2. Pussy Wagon
3. Sam Peckinpah
4. Charlton Heston
5. Eli Wallach
6. Jet Li
7. The Corruptor
8. Fistful of Dollars
9. The Searchers
10. Colin Farrell

ROUND 2
ACTION FILMS 16

QUESTIONS

1. Who played the Steve McQueen role in the remake of The Getaway?
2. In which film does a National Guard squad get lost in the Louisiana swamp?
3. Where do Stallone and others get trapped in Daylight?
4. What Stallone film has the tagline: 'Crime is the Disease. Meet the Cure'?
5. What actor plays Él Mariachi'?
6. What is the name of the sequel to Romancing the Stone?
7. Which TV series, also set in Hawaii, took the old slot of Hawaii Five-O?
8. In which war is Full Metal Jacket set?
9. In which city is Bullit set?
10. What film brought Paul Newman and Steve McQueen together for the second and last time?

ANSWERS

1. Alec Baldwin
2. Southern Comfort
3. Tunnel
4. Cobra
5. Antonio Banderas
6. Jewel of the Nile
7. Magnum PI
8. Vietnam
9. San Francisco
10. The Towering Inferno

ULTIMATE
BRAINBUSTER

ROUND 3

THRILLERS

QUESTIONS

1. On which famous American national monument does the final chase scene in North by Northwest take place?
2. What British actor plays the villain in North by Northwest?
3. Cary Grant worked with Hitchcock how many times?
4. What actor, often used by Hitchcock, starred in Rear Window?
5. What is Clint Eastwood's job in the film In the Line of Fire?
6. What was the profession at the centre of the drama Backdraft?
7. Which actor won an Oscar for his role as Deputy Marshal Samuel Gerard in The Fugitive?
8. What film follows Deputy Marshall Samuel Gerard on a pursuit of Wesley Snipes?
9. Who is Brad Pitt's partner in Se7en?
10. Who plays Bill Paxton's brother in A Simple Plan?

ANSWERS

1. Mount Rushmore
2. James Mason
3. Four
4. James Stewart
5. Secret Service Agent
6. Fire fighting
7. Tommy Lee Jones
8. U.S. Marshals
9. Morgan Freeman
10. Billy Bob Thornton

ROUND 3
THRILLERS 2

QUESTIONS

1. 'We're gonna need a bigger boat' is a line from what film?
2. In which film does Marion Crane die in the shower?
3. The Birds features which blonde actress in the lead?
4. What building does King Kong climb in the 1933 film?
5. 'She's my sister – my daughter' is a confession from which film?
6. The Manchurian Candidate (1960) stars which famous singer?
7. What does Robert Mitchum have tattooed on his knuckles in The Night of the Hunter?
8. In which city is Bullit set?
9. What does Grace Kelly use to kill the intruder in Dial M for Murder?
10. Who plays a Nazi dentist in The Marathon Man?

ANSWERS

1. Jaws
2. Psycho
3. Tippi Hedren
4. Empire State Building
5. Chinatown
6. Frank Sinatra
7. Love/Hate
8. San Francisco
9. Scissors
10. Lawrence Olivier

1. In which film does Audrey Hepburn play a blind woman?
2. Which actor plays Harry Lime in The Third Man?
3. The Thing was a remake of which 1951 film?
4. What film features Jane Fonda as a reporter at a nuclear plant accident?
5. Who plays the victimized wife in Gaslight (1944)?
6. Where do the Nazis store the uranium in Notorious?
7. Who does Matt Damon impersonate in The Talented Mr Ripley?
8. Where does Steven Spielberg's first film, Duel, take place?
9. What was Clint Eastwood's first film as a director?
10. In which film is Nicole Kidman terrorized on a boat by Billy Zane?

1. Wait Until Dark
2. Orson Welles
3. The Thing From Another World
4. The China Syndrome
5. Ingrid Bergman
6. Wine bottles
7. Dickie Greenleaf (Jude Law)
8. On the highway
9. Play Misty For Me
10. Dead Calm

ROUND 3
THRILLERS 4

QUESTIONS

1. Who plays Bridget Fonda's obsessive room-mate in Single White Female?
2. What is Alec Baldwin's profession in Malice?
3. What film did Kathy Bates win an Oscar for?
4. Who starred in The Eyes of Laura Mars?
5. Charade was remade in 2003 with which title?
6. John Travolta starred in a loose remake of the film Blow Up. What was it called?
7. What Tony Scott film features Gene Hackman as a surveillance expert?
8. What film tells the story of the failed assassination attempt on Charles de Gaulle?
9. What sporting event was the target of the terrorists in Black Sunday?
10. In which film is Robert Redford on the run from Max Von Sydow with Faye Dunaway?

ANSWERS

1. Jennifer Jason Leigh
2. Doctor
3. Misery
4. Faye Dunaway
5. The Truth About Charlie
6. Blow Out
7. Enemy of the State
8. The Day of the Jackal
9. Superbowl
10. Three Days of the Condor

1. Who plays killer John Doe in the film Se7en?
2. What thriller starring Guy Pearce is told backwards?
3. What 1949 Hitchcock film starring James Stewart contains no cuts?
4. In which Marilyn Monroe film does she plot the death of her husband?
5. Who played the Robert Mitchum part in the Cape Fear remake?
6. In Seven Days in May, which part does Burt Lancaster play?
7. Who stars in John Frankenheimer's thriller Seconds?
8. Who plays Secretary of Defence in the film No Way Out?
9. Who played Hannibal Lecter in Manhunter?
10. What is the name of Michael Caine's character in The Ipcress File?

1. Kevin Spacey
2. Memento
3. Rope
4. Niagara
5. Robert De Niro
6. General Scott
7. Rock Hudson
8. Gene Hackman
9. Brian Cox
10. Harry Palmer

QUESTIONS

1. Who stars in the paranoid conspiracy thriller Parallax View?
2. What film stars Julia Roberts and Mel Gibson?
3. What is Glenn Close's profession in Jagged Edge?
4. The Spy Who Came in From the Cold is based on a novel by which author?
5. Who are the journalists in the film All the President's Men?
6. Thirteen Days chronicles which true-life crisis from 1963?
7. Who plays British agent George Smiley?
8. What Henry Fonda film depicts an accidental nuclear war?
9. In which film does Sean Penn sell government secrets to the Soviet Union?
10. In Crimson Tide, Denzel Washington plays a Naval Officer rebelling against which Captain?

ANSWERS

1. Warren Beatty
2. Conspiracy Thriller
3. Lawyer
4. John Le Carre
5. Carl Bernstein and Bob Woodward
6. Cuban Missile Crisis
7. Alec Guinness
8. Fail-Safe
9. The Falcon and the Snowman
10. Gene Hackman

QUESTIONS

1. In how many films did Michael Caine feature as spy Harry Palmer?
2. Who directed Presumed Innocent, starring Harrison Ford?
3. What film pairs Denzel Washington with Julia Roberts?
4. In which film does Jane Fonda play a prostitute?
5. Consenting Adults features which Academy Award-winning actor as a conman and killer?
6. In which film does Jamie Foxx play a hijacked cab driver?
7. In which film does Mel Gibson play a disenchanted minister on a farm?
8. Bruce Willis stars opposite what actor in Unbreakable?
9. Which actress plays the babysitter in The Hand That Rocks the Cradle?
10. What film stars Uma Thurman, Richard Gere and Kim Basinger?

ANSWERS

1. Three – Ipcress File, Funeral in Berlin, Midnight in St Petersburg
2. Alan J. Pakula
3. The Pelican Brief
4. Klute
5. Kevin Spacey
6. Collateral
7. Signs
8. Samuel L. Jackson
9. Rebecca de Mornay
10. Final Analysis

ROUND 3
THRILLERS 8

QUESTIONS

1. What was James Stewart's profession in Rear Window?
2. Who won an Oscar for their performance in The Usual Suspects?
3. Who does Harrison Ford believe killed his wife in The Fugitive?
4. What does James Stewart believe to have witnessed through his Rear Window?
5. Gene Hackman stars in Enemy of the State alongside which famous rapper/actor?
6. What country does Michael Douglas travel to in Black Rain?
7. Who plays the President of the United States in Absolute Power?
8. John Cusack plays a juror for sale in which courtroom thriller?
9. In which city does Phone Booth take place?
10. Sandra Bullock plays an anti-death penalty activist in which John Grisham adaptation?

ANSWERS

1. Photographer
2. Kevin Spacey
3. The one-armed man
4. A murder
5. Will Smith
6. Japan
7. Gene Hackman
8. Runaway Jury
9. New York
10. A Time to Kill

1. In The Firm, what is Tom Cruise's profession?
2. Who plays the terrorizing husband in Gaslight (1944)?
3. What does the protagonist of Peeping Tom use to kill his victims?
4. Audrey Hepburn and which other actor starred in Charade?
5. Who plays Nicole Kidman's husband in Dead Calm?
6. What actress plays both sisters in the film Sisters (1973)?
7. In which film does Michael Caine play a transvestite killer?
8. Melanie Griffith plays a porn star in what thriller?
9. In The Recruit, Al Pacino hires which actor to become a CIA agent?
10. The Andromeda Strain is based on a novel by which author?

1. Lawyer
2. Charles Boyer
3. Tripod
4. Cary Grant
5. Sam Neill
6. Margot Kidder
7. Dressed to Kill
8. Body Double
9. Colin Farrell
10. Michael Crichton

QUESTIONS

1. Strangers on a Train featured a screenplay by which detective novelist?
2. What does Michael Caine cook to seduce his female colleague in The Ipcress File?
3. What is the nickname of Kevin Spacey's character in The Usual Suspects?
4. What does Glenn Close find to reveal that Jeff Bridges is the killer in Jagged Edge?
5. In which Bond movie does Telly Savalas play Blofeld?
6. In what country does Edward Fox have his forged passport made in Day of the Jackal?
7. Why is Goldeneye so called?
8. What leisure pursuit is followed by 'the dude' in The Big Lebowski?
9. Who plays the chief villain in North by Northwest?
10. The key to The 39 Steps is revealed by what sort of vaudeville performer?

ANSWERS

1. Raymond Chandler
2. An omelette
3. Verbal
4. A used typewriter ribbon
5. On Her Majesty's Secret Service
6. Italy
7. It was the name of Ian Fleming's Jamaican home
8. Tenpin Bowling
9. James Mason
10. A memory man

QUESTIONS

1. What thriller paired Michael Caine and Laurence Olivier?
2. What Superman actor stars in the film Deathtrap?
3. In which city is The Third Man set?
4. What suspense film stars Cher as a lawyer?
5. In which film is Cary Grant a murder suspect?
6. The Man Who Knew Too Much was filmed how many times by Hitchcock?
7. In which country did Hitchcock make The 39 Steps?
8. The Manchurian Candidate was remade with which actress in the Angela Lansbury role?
9. What was the name of The Wachowski Brothers' first film?
10. Al Pacino and which two-time Academy Award-winning actress star in Insomnia?

ANSWERS

1. Sleuth
2. Christopher Reeve
3. Vienna
4. Suspect
5. Suspicion
6. Twice
7. Britain
8. Meryl Streep
9. Bound
10. Hilary Swank

ROUND 3
THRILLERS 12

QUESTIONS

1. Pierce Brosnan stars as a millionaire art thief in which film?
2. Who is the female lead in The Talented Mr Ripley?
3. What is the name of Kenneth Branagh's first thriller?
4. The Life of David Gale stars which actor as an anti-death penalty activist?
5. What is Robin Williams' job in One Hour Photo?
6. Which President is Edward Fox trying to kill in The Day of the Jackal?
7. What area of France does To Catch a Thief take place in?
8. Which Danny De Vito comedy is a loose remake of Strangers on a Train?
9. What was the first time Michael Caine played Harry Palmer?
10. Sean Connery plays a Soviet Submarine captain in which film?

ANSWERS

1. The Thomas Crown Affair
2. Gwyneth Paltrow
3. Dead Again
4. Kevin Spacey
5. Photo store attendant
6. Charles De Gaulle
7. The Riviera
8. Throw Momma Off The Train
9. The Ipcress File
10. The Hunt for Red October

QUESTIONS

1. Which French hunk does Diane Lane cheat on Richard Gere with in Unfaithful?
2. In which film does Brittany Murphy say 'I'll Never Tell'?
3. In The Game, who buys Michael Douglas the Game for his birthday?
4. What is Guy Pearce pursuing in Memento?
5. Which John Le Carre adaptation stars Pierce Brosnan as a spy?
6. What disease does Sigourney Weaver suffer from in Copycat?
7. What British film stars Paddy Considine as a revenge-seeking killer?
8. What 1958 Orson Welles film stars Charlton Heston as a Mexican cop?
9. What famous director made the heist film The Killing in 1956?
10. In which city does Vertigo take place?

ANSWERS

1. Olivier Martinez
2. Don't Say A Word
3. Sean Penn – his brother
4. The man who killed his wife
5. The Tailor of Panama
6. Agoraphobia
7. Dead Man's Shoes
8. Touch of Evil
9. Stanley Kubrick
10. San Francisco

QUESTIONS

1. Which author wrote the books on which these films are based: Pelican Brief, The Firm, and A Time to Kill?
2. In which film does Susan Sarandon play a lawyer representing a child witness?
3. In which 1998 thriller set in L.A. is there a bordello with prostitutes that look like movie stars?
4. The director of the Usual Suspects went on to direct a popular superhero film – which one?
5. Who stars as a Texas Ranger in the film Extreme Prejudice?
6. In which film does Jamie Lee Curtis play a rookie cop?
7. Who plays Woody Harrelson's brother in the film Money Train?
8. In which film does Steven Seagal play an Italian Brooklyn cop?
9. Which author is responsible for creating the Hannibal Lecter character?
10. Who stars in Don't Say a Word?

ANSWERS

1. John Grisham
2. The Client
3. L.A. Confidential
4. X-Men
5. Nick Nolte
6. Blue Steel
7. Wesley Snipes
8. Out for Justice
9. Thomas Harris
10. Michael Douglas

QUESTIONS

1. Which actor played Hannibal Lecter in Manhunter?
2. In which film are the seven deadly sins a serial killer's motive?
3. Who plays Al Pacino's partner in the film Sea of Love?
4. Who stars in Three Days of the Condor?
5. During whose American Presidency is the film Thirteen Days set?
6. Who plays Lee Harvey Oswald in the film JFK?
7. Vanessa Redgrave plays the arms dealer villain Max in which film?
8. In which film does Harrison Ford foil an IRA assassination plot?
9. In which country does the beginning of The Bourne Supremacy take place?
10. Which espionage film pairs Brad Pitt and Robert Redford?

ANSWERS

1. Brian Cox
2. Se7en
3. John Goodman
4. Robert Redford
5. John F. Kennedy
6. Gary Oldman
7. Mission: Impossible
8. Patriot Games
9. India
10. Spy Games

QUESTIONS

1. Which David Mamet film has Val Kilmer rescuing the President's daughter?
2. What serial killer film stars Angelina Jolie and Denzel Washington?
3. Who stars in The General's Daughter?
4. Which character actress plays Jimmy Stewart's wise-cracking nurse in Rear Window?
5. In the Thomas Crown Affair remake, which part does Faye Dunaway play?
6. Who does Johnny Depp pursue in the film From Hell?
7. In which film does Keanu Reeves work at the Devil's law firm?
8. What John Carpenter film features a group of cops and criminals trapped in a police station?
9. In Ronin, who plays Robert De Niro's French partner?
10. In which film is Julianne Moore convinced, against all evidence, that she had a son?

ANSWERS

1. Spartan
2. The Bone Collector
3. John Travolta
4. Thelma Ritter
5. Pierce Brosnan's psychiatrist
6. Jack the Ripper
7. Devil's Advocate
8. Assault on Precinct 13
9. Jean Reno
10. The Forgotten

ROUND 4
HORROR

1.   Which author wrote the original Frankenstein novel?
2.   How many sequels to The Exorcist have been made?
3.   What was the name of Shelley Duvall's character in The Shining?
4.   Which actor played the monster in the original film of Frankenstein?
5.   Who directed the original version of The Haunting?
6.   Mel Brooks directed a spoof of Frankenstein in 1974 – what's the title?
7.   How many sequels to Hellraiser have there been?
8.   What did the film-makers use to imitate vomit in The Exorcist?
9.   Which actress played Clarice Sterling in Hannibal?
10. Who is the director of Amityville?

1.   Mary Shelly
2.   Three
3.   Wendy Torrance
4.   Boris Karloff
5.   Robert Wise
6.   The Young Frankenstein
7.   Eight
8.   Pea Soup
9.   Julianne Moore
10. Stuart Rosenberg

ROUND 4
HORROR 2

QUESTIONS

1. In 1999 The Haunting was released, a remake of a film of the same name released in what decade?
2. Who played the monster in Kenneth Branagh's Frankenstein?
3. What is the name of the villain in the Hellraiser films?
4. In which Vincent Price film does he offer $10,000 to anyone who can survive the night?
5. What is the name of the actor who plays The Exorcist in the 1973 film?
6. In Hannibal, which part of Ray Liotta's body does Anthony Hopkins cook?
7. What is the name of the hotel in The Shining?
8. What was the name of the sequel made four years after the original Frankenstein?
9. What is the name of the evil child in The Omen?
10. In The Shining, what does Danny see a pair of in the corridor of the hotel?

ANSWERS

1. 1960s (1963)
2. Robert De Niro
3. Pinhead
4. House on Haunted Hill
5. Max Von Sydow
6. Brain
7. Overlook Hotel
8. The Bride of Frankenstein
9. Damien
10. Identical twins

QUESTIONS

1. In Red Dragon, which actor plays serial killer The Toothfairy?
2. How did Hannibal Lecter cook a census taker's liver in The Silence of the Lambs?
3. What does REDRUM mean in The Shining?
4. What was the first Stephen King book to be turned into a film?
5. How does Candyman appear?
6. The Blair Witch Project was released in which year?
7. What does Kathy Bates force James Caan to do in Misery?
8. Who directed The Night of The Living Dead?
9. What film used outtakes from The Shining for its final scene?
10. What is Jamie Lee Curtis' relation to the killer in Halloween?

ANSWERS

1. Ralph Fiennes
2. 'With some fava beans and a nice Chianti.'
3. MURDER
4. Carrie
5. By saying his name three times
6. 1999
7. Write a book
8. George Romero
9. Blade Runner
10. Sister

QUESTIONS

1. What was the name of the creator of The Twilight Zone?
2. In which Nightmare on Elm Street film does Robert Englund (Freddy Krueger) play himself?
3. Which British director made 28 Days Later?
4. Which child actor stars opposite Bruce Willis in The Sixth Sense?
5. Which horror film featured Jamie Lee Curtis and her mother Janet Leigh together for the first time?
6. Who composed the music for the film Halloween?
7. What weapon does Freddy Kreuger use to kill?
8. What is the name of Buffy's High School?
9. The Dead Zone is based on a novel by which writer?
10. What is the most successful horror film of all time?

ANSWERS

5. The Fog
4. Haley Joel Osment
3. Danny Boyle
2. New Nightmare
1. Rod Serling

10. The Sixth Sense
9. Stephen King
8. Sunnydale High School
7. Glove with finger knives
6. John Carpenter

QUESTIONS

1. Who played Count Dracula in Francis Ford Coppola's Bram Stoker's Dracula?
2. Who wrote the book on which the film Interview with the Vampire is based?
3. In which southern American city is the film Dracula 2000 based?
4. John Carpenter directed a Stephen King story about a killer car – what is it called?
5. What actor played the original Dracula in 1931?
6. Who stars as Buffy in the TV series?
7. What film features John Goodman as a bug exterminator?
8. In which city does The Sixth Sense take place?
9. According to the Guinness Book of Records, which horror character is the most frequently played in films?
10. In which town does Buffy the Vampire Slayer take place?

ANSWERS

1. Gary Oldman
2. Anne Rice
3. New Orleans
4. Christine
5. Boris Karloff
6. Sarah Michelle Gellar
7. Arachnophobia
8. Philadelphia
9. Dracula
10. Sunnydale

QUESTIONS

1. What John Carpenter film stars Donald Pleasance as a Priest?
2. What was the latest Exorcist sequel to have been made?
3. What actor played The Phantom of the Opera in the original film?
4. In what film does Julia Roberts play Dr. Jekyl's maid?
5. In what 3-D horror film does Vincent Price turn his enemies into wax figures?
6. What was Francis Ford Coppola's early horror film called?
7. What was the only Roman Polanski film to star his late wife Sharon Tate?
8. What is the name of a 1985 Tobe Hooper film in which London is overrun with zombies?
9. Who stars as Wendy Torrance in The Shining?
10. What Hugh Jackman film stars most of Universal's horror legends?

ANSWERS

1. Prince of Darkness
2. Exorcist: The Beginning
3. Lon Chaney
4. Mrs. Reily
5. House of Wax
6. Dementia 13
7. The Fearless Vampire Killers
8. Lifeforce
9. Shelly Duvall
10. Van Helsing

1. Which zombie film, adapted from a video game, stars Milla Jovovich?
2. In Duel, what kind of vehicle terrorizes Dennis Weaver?
3. Which pop star starred in Videodrome?
4. Where does Jason X take place?
5. Which actor played Darkman?
6. Who stars as Ichabod Crane in Sleepy Hollow?
7. What allows Gremlins to reproduce?
8. The Craft and Scream share which actress in common?
9. What is the name of the second sequel to Evil Dead?
10. What's the name of the chainsaw-wielding killer in Texas Chainsaw Massacre?

ANSWERS

1. Resident Evil
2. A truck
3. Debbie Harry
4. Outer Space
5. Liam Neeson
6. Johnny Depp
7. Water
8. Neve Campbell
9. Army of Darkness
10. Leatherface

QUESTIONS

1. How many sequels to Psycho were made?
2. Where does Jaws take place?
3. Where does Jack Nicholson eventually die at the end of The Shining?
4. How many times has the Invasion of the Body Snatchers been remade?
5. Where does The Thing take place?
6. How many films have John Carpenter and Kurt Russell made together?
7. The Birds is based on a story by which writer?
8. In which Roman Polanksi film does Catherine Deneuve lose her mind?
9. Which Spanish filmmaker directed The Others?
10. What phrase does Jack Nicholson write endlessly on his typewriter in The Shining?

ANSWERS

1. Three
2. Amity Island
3. In the maze
4. Twice
5. Antarctica
6. Five
7. Daphne Du Marnier
8. Repulsion
9. Alejandro Amenabar
10. 'All work and no play make Jack a dull boy.'

1. When Jack Nicholson rips apart the bathroom door in The Shining, what does he say?
2. What is Ellen Burstyn's profession in The Exorcist?
3. In The Omen, what sign indicates that Damien is the antichrist?
4. In Halloween, what is the name of the killer?
5. In 1999, The House on Haunted Hill was remade – who played the Vincent Price role?
6. In what film does Mia Farrow give birth to Satan's son?
7. What was Stanley Kubrick's next film after The Shining?
8. Which film features a headless horseman?
9. What film had the tagline – 'Keep telling yourself: It's only a movie'?
10. Who plays the Marine Biologist in Jaws?

1. Here's Johnny!
2. Film actress
3. 666 on his scalp
4. Michael Myers
5. Geoffrey Rush
6. Rosemary's Baby
7. Full Metal Jacket
8. Sleepy Hollow
9. Last House on the Left
10. Richard Dreyfuss

QUESTIONS

1. Poltergeist was produced by which famous director?
2. James Whale directed Frankenstein (1931). He also directed an H.G. Welles adaptation starring Claude Rains. Which film is this?
3. What 1932 Tod Browning film takes place in a circus?
4. What is the name of the villain who appears in dreams in the Nightmare on Elm Street films?
5. Dawn of the Dead is a sequel to which 1968 horror classic?
6. In which Friday 13th film does Jason Voorhees make his first appearance?
7. Where does most of the action take place in The Dawn of the Dead?
8. What Simon Pegg 2003 comedy film is a spoof of The Dawn of the Dead and many other zombie films?
9. The director of Spiderman also wrote and directed the Evil Dead films. What's his name?
10. On what grounds was the house in Poltergeist built?

ANSWERS

1. Steven Spielberg
2. The Invisible Man
3. Freaks
4. Freddy Krueger
5. Night of the Living Dead Part II
6. Part II
7. Day of the Dead
8. A shopping mall
9. Sam Raimi
10. Indian burial grounds

1. What was Peter Jackson's first film?
2. In which film does Kate Winslet have an imaginary life that spills out of control into murder?
3. What British film company was known for films such as 'The Quartermass Experiment', 'Brides of Dracula' and for using actors Christopher Lee and Peter Cushing?
4. Which role was Christopher Lee most famous for?
5. Which Academy-Award nominee and sex symbol appears in the first Nightmare on Elm Street?
6. What kind of mask does Jason Voorhees wear in the Friday the 13th films?
7. What film has the tagline 'When there is no more room in hell, the dead will walk the earth'?
8. Kevin Bacon makes an appearance in which Friday the 13th film?
9. Jason Vorhooees is pitted against another legendary villain in which 2003 film?
10. What is the name of the sequel to Dawn of the Dead?

1. Brain Dead
2. Heavenly Creatures
3. Hammer
4. Dracula
5. Johnny Depp
6. Hockey Mask
7. Dawn of the Dead
8. Friday the 13th I
9. Freddy Vs. Jason
10. Day of the Dead

QUESTIONS

1. What silent German film directed by F.W. Murnau was one of the early vampire films?
2. Who played the lead in the original Phantom of the Opera?
3. How many sequels did the Texas Chainsaw Massacre spawn?
4. What power does Carrie possess?
5. Before being a long-running series, Buffy the Vampire Slayer was a film starring whom?
6. Where was the remake of An American Werewolf in London set?
7. What actor stars in The Mummy (1999)?
8. What character have both Anthony Hopkins and Hugh Jackman played on screen?
9. Whicih actor plays the vampire killer Blade?
10. What secret does the child have in The Sixth Sense?

ANSWERS

1. Nosferatu
2. Lon Chaney
3. Four
4. Telekinesis
5. Kristy Swanson
6. Paris
7. Brendan Fraser
8. Van Helsing
9. Wesley Snipes
10. He sees dead people

1. In what transport system does Death Line take place?
2. What was Ealing Studio's first foray into horror?
3. In which 1961 film does Deborah Kerr play a governess?
4. Who starred in the remake of The Texas Chainsaw Massacre?
5. Who starred in and directed the 1994 film Mary Shelley's Frankenstein?
6. Who plays the title role in the film Young Frankenstein?
7. Asia Argento is the daughter of which Italian horror director?
8. In which film does James Woods star as a Vampire hunter?
9. Which David Cronenberg film stars Oliver Reed?
10. What British horror film is set in a German trench in World War I?

1. London Underground
2. Dead of Night
3. The Innocents
4. Jessica Biel
5. Kenneth Brannagh
6. Gene Wilder
7. Dario Argento
8. Vampires
9. The Brood
10. Deathwatch

QUESTIONS

1. Tobe Hooper, director of The Texas Chainsaw Massacre also directed an adaptation of a Stephen King novel – what was it called?
2. Which British director made 28 Days Later?
3. Who played Sgt. Neil Howie in The Wicker Man?
4. What actor starred as Dr Phibes?
5. What actor plays a villain in the film 10 Rillington Place?
6. Which Irish filmmaker directed The Company of Wolves?
7. Interview with the Vampire stars what actor as Lestat?
8. Which British horror film had the tagline 'Six Soldiers. Full Moon. No chance'?
9. In which city does Donald Sutherland die in Don't Look Now?
10. What comes pouring out of the lift doors in The Shining?

ANSWERS

1. Salem's Lot
2. Danny Boyle
3. Edward Woodward
4. Vincent Price
5. Richard Attenborough
6. Neil Jordan
7. Tom Cruise
8. Dog Soldiers
9. Venice
10. Blood

QUESTIONS

1. What horror film did Rob Zombie direct?
2. In Bride of Chucky, who plays the evil doll's wife?
3. Which Halloween film marked the return of Jamie Lee Curtis to the series after almost 20 years?
4. Drew Barrymore stars as a child with powers in which Stephen King adaptation?
5. In which film does a little girl say 'They're here' while pointing at the TV?
6. The Village of the Damned was remade in the 90s by which famous horror director?
7. What is the name of the HBO TV series featuring short horror stories?
8. Which actress gets killed in the first fifteen minutes of Scream?
9. What was the sequel to The Blair Witch project called?
10. Who plays George Clooney's brother in From Dusk Till Dawn?

ANSWERS

1. House of 1000 Corpses
2. Jennifer Tilly
3. Halloween H20
4. Firestarter
5. Poltergeist
6. John Carpenter
7. Tales From The Crypt
8. Drew Barrymore
9. Book of Shadows: Blair Witch 2
10. Quentin Tarantino

ROUND 4
HORROR 16

QUESTIONS

1. What film had the tag line "In space no-one can hear you scream"?
2. In which film does Oscar winner Charlize Theron star opposite Keanu Reeves?
3. What is the name of the film directed by Roman Polanski and starring Johnny Depp?
4. In which film do the monsters from the Predator and Alien films meet?
5. The Children of the Corn is based on a novel by which horror author?
6. In which film does Kiefer Sutherland play a vampire?
7. What horror film stars Sarah Michelle Gellar as a beauty queen murdered by a man with a hook?
8. What was the name of the first film featuring evil doll Chucky?
9. Who stars in Peter Jackson's The Frighteners?
10. Which Friends actress stars as a ruthless reporter in Scream?

ANSWERS

1. Alien
2. Devil's Advocate
3. The Ninth Gate.
4. Alien Vs. Predator
5. Stephen King
6. The Lost Boys
7. I Know What you Did Last Summer
8. Child's Play
9. Michael J. Fox
10. Courtney Cox

1. What 1972 film tells the story of a country trip gone terribly wrong?
2. In what film does Jack Nicholson become a werewolf?
3. What was the killer in the film It's Alive?
4. What is the name of the 1995 Texas Chainsaw Massacre sequel starring Matthew McConaughey and Renee Zellweger?
5. Who plays the title role in Carrie?
6. Who plays the Devil in the film End of Days?
7. What Michael J. Fox film has him as a werewolf?
8. What 1983 film, based on a Stephen King book, features a dog with rabies?
9. What is the profession of Fay Wray`s character in the 1933 film King Kong - Journalist, Writer or Actress?
10. Who plays the Devil in The Devil's Advocate?

1. Deliverance
2. Wolf
3. Baby
4. The Next Generation
5. Sissy Spacek
6. Gabriel Byrne
7. Teen Wolf
8. Cujo
9. Actress
10. Al Pacino

QUESTIONS

1. Which Edgar Allan Poe film adaptation features Boris Karloff, Vincent Price and Jack Nicholson?

2. What cheap B horror movie from 1958 featured a young Steve McQueen?

3. Bela Lugosi and Boris Karloff starred in three films together. Name the first one?

4. The Hammer 1959 Mummy film featured which actor as the mummy?

5. What giant insects attack in Them!?

6. What was Charles Laughton's only directorial effort?

7. Which composer scored Psycho?

8. Roman Polanski's wife Sharon Tate died at the hands of which notorious cult leader's gang?

9. What kind of films is director George Romero best known for?

10. Who stars in Don't Look Now alongside Donald Sutherland?

ANSWERS

1. The Raven
2. The Blob
3. The Black Cat
4. Christopher Lee
5. Ants
6. Night of the Hunter
7. Bernard Hermann
8. Charles Manson
9. Zombie films
10. Julie Christie

ULTIMATE
BRAINBUSTER

ROUND 5
CHILDREN

S FILMS

1. What was Mickey Mouse's screen debut?
2. What cartoon character needs spinach to gain strength?
3. What cartoon duet of cat and mouse endlessly battle each other?
4. What British studios produce the claymation adventures of Wallace & Grommit?
5. Who is the author of the Harry Potter books?
6. In what Spielberg-produced film do a gang of kids go hunting for a pirate's treasure?
7. In which films does Macaulay Culkin get left alone?
8. What two animated films about insects were released in 1998?
9. What is the name of a popular large purple dinosaur?
10. What name is normally given to Japanese Animation?

1. Steamboat Willie
2. Popeye
3. Tom & Jerry
4. Aardman Studios
5. J.K. Rowling
6. The Goonies
7. Home Alone 1 & 2
8. Antz and Bugs' Life
9. Barney
10. Anime

ROUND 5
CHILDREN'S FILMS 2

1. What was the name of Samantha and Darrin's daughter on Bewitched
2. What was the first Lassie film?
3. What was the name of Herman's wife on The Munsters?
4. What kind of animal is Babe?
5. Where did Huckleberry Hound live?
6. Who lives next door to the Flintstones?
7. What animal was at the centre of Free Willy?
8. Which Wacky Racers drive the Bouldermobile?
9. Natalie Wood stars in which Christmas Classic about Santa Claus?
10. What was the name of the first Muppet film?

10. The Muppet Movie
9. Miracle on 34th Street
8. The Slag Brothers
7. Killer Whale
6. Barney, Betty and Bam Bam Rubble

5. Jellystone park
4. Pig
3. Lily
2. Lassie Come Home
1. Tabitha

1. What was the name of Lady Penelope's chauffeur in Thunderbirds?
2. Who directed the first two Harry Potter films?
3. What was Basil Brush's catchphrase?
4. In which Tim Burton film is Santa Clause captured and tortured?
5. Which ex-Goon was the star of 'Potty Time'?
6. The Lion, The Witch and the Wardrobe is part of which Chronicles?
7. In which fictional town did Larry the Lamb and his friends live?
8. Who plays Hazel in Watership Down?
9. What was the name of the two sensational singing piglets whose TV show ran from the late 50s to the early 70s?
10. What animated film tells the story of a widowed mouse seeking help from rats?

1. Parker
2. Chris Columbus
3. Boom! Boom!
4. Nightmare Before Christmas
5. Michael Bentine
6. of Narnia
7. Toytown
8. John Hurt
9. Pinky and Perky
10. The Secret of NIMH

ROUND 5
CHILDREN'S FILMS 4

1. Who did Andy Pandy live with in his basket?
2. Who played Doctor Dolittle in the 1967 version?
3. In The Herbs, what was the name of the lion?
4. In which film does the song 'It's A Hard Knock Life' feature?
5. What was the name of the wooden woodpecker bookend in Bagpuss?
6. In which film does Robin Williams dress up as a woman in order to see his children?
7. Who ran the fort in Camberwick Green?
8. In which Jim Henson film does a Gelfin named Jen topple the evil Skeksis?
9. Who was the first Blue Peter dog?
10. What Christmas animated film stars Tom Hanks in six different roles?

5. Professor Yaffle
4. Annie
3. Parsley
2. Rex Harrison
1. Looby Loo and Teddy

10. The Polar Express
9. Petra
8. Dark Crystal
7. Captain Snort
6. Mrs. Doubtfire

1. What was the name of Honk Kong Phooey's faithful cat?
2. Who starred in the remake of Freaky Friday?
3. Which Simpsons character has a roof top garden that Paul and Linda McCartney occasionally visit?
4. Who starred in the original version of The Parent Trap?
5. In The Flumps, what was the name of Perkin and Posy's younger brother?
6. What was the currency used in The Adventure Game?
7. Who starred in the original Freaky Friday as the young girl?
8. Which Saturday morning show invented the 'Dying Fly' dance?
9. What kind of dog was Beethoven?
10. What was the name of the Jim Henson film starring David Bowie and Jennifer Connelly?

1. Spot
2. Lindsay Lohan
3. Apu Nahasapeemapetilon
4. Hayley Mills
5. Pootle
6. Drogna
7. Jodie Foster
8. Tiswas
9. St. Bernard
10. Labyrinth

ROUND 5
CHILDREN'S FILMS 6

QUESTIONS

1. What animation studio in association with Disney produced the films Toy Story and Finding Nemo?
2. What's the name of the heroic collie starring in so many films and TV series?
3. What film tells the story of a child and his black horse?
4. Who created The Muppets?
5. Who wrote the book on which Willy Wonka and the Chocolate Factory is based?
6. What breed of dog was Rin Tin Tin?
7. What was the sequel to Babe called?
8. What green monster is voiced by Mike Myers?
9. What story about a boy that never grows was written by J.M. Barrie?
10. In what decade did Blue Peter start broadcasting?

ANSWERS

10. 1950s (1958)
9. Peter Pan
8. Shrek
7. Babe: Pig in the City
6. German Shepherd

5. Road Dahl
4. Jim Henson
3. Black Stallion
2. Lassie
1. Pixar

1. What popular comic book featuring a young Belgian journalist was later turned into a series of animated films?
2. What actress plays the title character in Mary Poppins?
3. What was the first film to combine animated characters with live-action film?
4. In The Muppet Christmas Carol, who played Scrooge?
5. Who created The Magic Roundabout?
6. What studio produced the Looney Tunes?
7. What was Disney's first full length animated film?
8. Which was Mickey Mouse's only appearance in a full-length film?
9. The comic strip and cartoons Peanuts featured which famous beagle?
10. What is the name of the owner of the nuclear plant in The Simpsons?

10. Mr. Burns
9. Snoopy
8. Fantasia
7. Snow White and the Seven Dwarfs
6. Warner Brothers

5. Eric Thompson
4. Michael Caine
3. Anchors Aweigh
2. Julie Andrews
1. The Adventures of Tintin

ROUND 5
CHILDREN'S FILMS 8

1. What show announced itself as 'It's Friday, it's five to five, it's...'
2. Who created the Simpsons?
3. What film quiz did Michael Rodd host?
4. What brand of cigarettes did The Flinstones endorse at the beginning of their run?
5. The Bay City Rollers had their own TV show, what was it called?
6. What film tells the story of a small bear cub trying to survive after his mother dies?
7. What was the name of the Banana Splits rival gang?
8. Frankie Muniz stars in which film about a terrier in Mississippi?
9. Which teen heart-throb played Keith in The Partridge Family?
10. Who is Miss Piggy in love with?

1. Crackerjack
2. Matt Groening
3. Screen Test
4. Winstons
5. Shang-a-lang
6. The Bear
7. The Sour Grapes Gang
8. My Dog Skip
9. David Cassidy
10. Kermit

1. Which animal show did Clarence the cross-eyed lion star in?
2. What studio produced The Wizard of Oz?
3. Helicopter pilots Chuck Martin and PT Moore featured in which show?
4. What was the name of Jon Petwee's Dr Who's yellow car?
5. In The Simpsons what is the name of Selma Bouvier's pet iguana?
6. Where does Home Alone 2 take place?
7. What was the name of the German-imported show which featured the arrogant Princess Thousandbeauty?
8. What children's film features Elizabeth Taylor and horse-races?
9. What actress played the twins in the original of The Parent Trap?
10. What was the name of the child who takes in the alien in E.T.?

1. Daktari
2. MGM
3. Whirlybirds
4. Bessie
5. Jub Jub
6. New York
7. The Singing Ringing Tree
8. National Velvet
9. Hayley Mills
10. Elliot

ROUND 5
CHILDREN'S FILMS 10

1. What kind of creature was Mungo in 'Mary, Mungo and Midge'?
2. In which city does Stuart Little live?
3. What was the name of the family in Button Moon?
4. Who plays Prince Charming in Shrek?
5. Who is the object of Worzel Gummidge's affections?
6. Who plays Rocky the chicken in Chicken Run?
7. Who runs the kitchen in The Wombles?
8. What is the name of the cabin boy in Captain Pugwash?
9. What is the name of the actor who plays Harry Potter?
10. What Spanish actor is the young spies' father in Spy Kids?

1. He was a dog
2. New York
3. The Spoon family
4. Rupert Everret
5. Aunt Sally
6. Mel Gibson
7. Madame Cholet
8. Tom
9. Daniel Radcliffe
10. Antonio Banderas

QUESTIONS

1. Tiger, Sticks, Doughnut, Billie, Spring, Scooper and Brains appeared in which show?
2. What is the name of the character voiced by Tom Hanks in Toy Story 1 & 2?
3. Which show advised you to 'Switch off your television set and go and do something less boring instead…?'
4. Where does Nemo get put after being captured?
5. Rodney Bennett and Brookdale appear in which school soap?
6. What film features Michael Jordan and the Looney Tunes?
7. Who owns a black and white cat called Jess?
8. Bella, Milo, Fizz and Jake appear on which show?
9. In which country is The Secret of Roan Inish set?
10. What classic children's film was made in both 1949 and 1993 and is based on a novel by Frances Hogdson Burnett?

ANSWERS

1. Here Comes The Double Deck-
ers
2. Woody
3. Why Don't You?
4. A fish tank
5. Grange Hill
6. Space Jam
7. Postman Pat
8. Tweenies
9. Ireland
10. The Secret Garden

ROUND 5
CHILDREN'S FILMS 12

1. What was the name of the dragon that lived in Ivor the Engine's boiler?
2. On which road did Mr Benn live?
3. Martin Scorsese plays a shark called Sykes in which film?
4. What does Ralph Macchio learn from Mr. Miyagi so he can defend himself?
5. Which cartoon bears lived in Cave Block 9 of the Wonderland Zoo?
6. In which film does Rick Moranis shrink his children?
7. What was the name of Dangermouse's nemesis?
8. In which film do The Muppets travel to New York?
9. Who likes to sing, 'Can we fix it?'
10. What identical girl twins have starred in a series of TV shows and films together?

1. Idris
2. Festive Road, at number 52.
3. Shark Tale
4. Karate
5. The Hair Bear Bunch
6. Honey, I Shrunk The Kids
7. Baron Greenback
8. Muppets Take Manhattan
9. Bob the Builder
10. The Olsen Twins

1. What was the name of the sock puppet that Shari Lewis made famous?
2. What was the name of the first Harry Potter film?
3. What was the colour of Rod Hull and Emu's Windmill?
4. What Disney film is now an acclaimed stage production?
5. Which puppet's magic spell was 'Izzy whizzy, let's get busy'?
6. What animated film is the highest grossing Japanese film ever?
7. What was the name of the first pair of Blue Peter presenters?
8. What was the name of the hapless caretaker in Take Hart?
9. What lazy cat featured in a comic strip was turned into a film in 2004?
10. Who does Hayley Mills think she's found in her barn in Whistle Down The Wind?

1. Lambchop
2. Harry Potter and the Philosopher's Stone
3. Pink
4. Lion King
5. Sooty
6. Princess Mononoke
7. Christopher Trace and Leila Williams
8. Mr Bennet
9. Garfield
10. Jesus

QUESTIONS

1. Davy Jones, Mickey Dolenz, Mike Nesmith: who is the missing Monkee?
2. In The Muppet Movie, who sings 'It's Not Easy Being Green'?
3. Who presented and provided voices for all the animals in Animal Magic?
4. In which film does Jack Black teach a classroom how to rock?
5. The Soup Dragon and the Iron Chicken were characters on which animated show?
6. When was the first Winnie the Pooh film made?
7. In Hector's House, which cat does Kiki The Frog live next door to?
8. Where does Spongebob Squarepants live?
9. On The Simpsons, what is the name of the Comic Book Guy's shop?
10. What cartoon series had 'robots in disguise'?

ANSWERS

1. Peter Tork
2. Kermit the Frog
3. Johnny Morris
4. School of Rock
5. The Clangers
6. 1977
7. Zaza
8. Bottom of the sea
9. The Android's Dungeon
10. Transformers

QUESTIONS

1. Annette Mills sang with which puppet?
2. A Little Princess is based on a novel by the same author of The Secret Garden–what's her name?
3. On Fingerbobs, who was it who 'lifts a finger and a mouse is there...'
4. What characters in Willy Wonka and the Chocolate Factory sing whenever a child has misbehaved?
5. What was the name of the Roundabout owner in The Magic Roundabout?
6. What is the name of the creator of The Grinch?
7. What is the name of the Japanese studio responsible for Spirited Away?
8. Mr Meaker, Hazel the McWitch and Timothy Claypole appeared in which show?
9. Julia Sawalha and Dexter Fletcher appeared in which award winning show surrounding the Junior Gazette?
10. Who plays the Donkey in Shrek?

ANSWERS

1. Muffin the Mule
2. Frances Hogson Burnett
3. Yoffy
4. Oompah Loompah
5. Mr Rusty
6. Dr. Seuss
7. Studio Gibli
8. Rentaghost
9. Press Gang
10. Eddie Murphy

ROUND 5
CHILDREN'S FILMS 16

QUESTIONS

1. What is the name of Jamie's canine companion in Jamie and the Magic Torch?
2. Who plays the genie in Aladdin?
3. Which Channel 4 puppet caused parents to complain that he was encouraging children to spit?
4. How do the monsters generate electric power in Monsters Inc.?
5. Which TV show set in Newcastle first featured Ant and Dec?
6. In Harvey, what does James Stewart see that no one else can?
7. Which Medieval magician had a friend called Carrot?
8. Who plays Lucius Best in The Incredibles?
9. King Otto and Queen Doris were ruling monarchs in which animated show?
10. What film features the adventures of a cub bear after his mother dies?

ANSWERS

1. Wordsworth
2. Robin Williams
3. Pob
4. By scaring children
5. Byker Grove
6. A rabbit
7. Catweazle
8. Samuel Jackson
9. Chorlton and The Wheelies
10. The Bear

ROUND 5
CHILDREN'S FILMS 17

QUESTIONS

1. Which Saturday morning kids' show features Posh Paws and Brown Sauce?
2. What TV series and films feature four speaking turtles fighting for justice?
3. On which show did Spotty Dog appear?
4. In which film does a child find himself (literally) absorbed into the magical book he is reading?
5. In The Simpsons, what is the name of the doll that Smithers collects?
6. What was the only animated film to have ever received a Best Picture Nomination?
7. On The Magic Roundabout whose catchphrase was 'Time for bed…'
8. Where is Thunderbirds' base?
9. Before becoming a feature film, the Railway Children was a series on which TV channel?
10. What Disney film features the song 'Under the Sea'

ANSWERS

5. Malibu Stacy
4. The Neverending Story
3. The Woodentops
2. Teenage Mutant Ninja Turtles
1. The Multi Coloured Swap Shop

10. Little Mermaid
9. BBC
8. Tracy Island
7. Zebedee
6. Beauty and the Beast

ROUND 5
CHILDREN'S FILMS 18

1. Where does Paddington's Aunt Lucy live?
2. Alice in Wonderland is based on a story by which author?
3. What was the name of Rod and Freddy's female singing companion on Rainbow?
4. In which country does the Disney film Mulan take place?
5. What is the name of the police officer in Top Cat?
6. In which cartoon did the Teen Angels appear?
7. The 1967 Disney film Jungle Book is based on fiction by which author?
8. What is the name of Dick Dastardley's sniggering side-kick?
9. Who is the star of Chitty Chitty Bang Bang?
10. Doctor Dolittle was remade in 1998 with which actor?

1. A home for retired bears in Lima, Peru.
2. Lewis Carroll
3. Jane
4. China
5. Officer Dibble
6. Captain Caveman
7. Rudyard Kipling
8. Muttley
9. Dick Van Dyke
10. Eddie Murphy

ROUND 6

SOAP

PERAS

1. What was the name of the bar in Eldorado?
2. On the O.C what is the name of Ryan's ex-girlfriend who claimed to be pregnant?
3. Actors from which soap had a hit with the song 'Hillbilly Rock'?
4. What was the name of Heather Locklear's character in Dynasty?
5. What is the name of Pacey's sister who has a romance with Dawson from Dawson's Creek?
6. In Coronation Street, what is the nickname of Emily Bishop's nephew Geoffrey?
7. Who came back from the dead with the words, "'Ello Princess?"
8. What was Brookside's Peter Harrison once accused of?
9. What was the name of Kim Tate's horse that was killed during the Emmerdale plane crash?
10. Who was Home & Away's Roxy's half-brother?

9. Dark Star
8. Date rape
7. Eastenders' Den Watts aka actor Leslie Grantham
6. Spider
5. Gretchen
4. Sammy-Jo Dean Reece Carrington Fallmont
3. Emmerdale
2. Theresa
1. Joy's bar.

ROUND 6
SOAP OPERAS 2

QUESTIONS

1. Joan Collins played which character in Dynasty? Krystle, Fallon or Alexis?
2. What was the name of Miss Ellie's second husband on Dallas?
3. What disease did Eastenders' Mark Fowler discover he was carrying?
4. Which Coronation Street character became famous for her dangling ear rings, leopard-skin clothing and bleached blonde beehive?
5. Who runs the general store in Home & Away?
6. Who married Kirk Benson in Hollyoaks?
7. What pet did Phoebe keep in Neighbours? A dog, a bird or a snake?
8. What is the name of Bev McLoughlin's son on Brookside?
9. On Family Affairs, what is the name of Dave Callan's pub?
10. What illegal countryside activity are the Dingle family infamous for on Emmerdale?

ANSWERS

1. Alexis
2. Clayton Farlow
3. H.I.V.
4. Bet Lynch
5. Alf Stewart
6. Ruth Osborne
7. A snake called Oscar
8. Josh
9. The Black Swan
10. Poaching

1. Aaron Spelling's daughter appeared in which U.S teen soap?
2. What job did Top Dog Bea Smith do in Prisoner Cell Block H?
3. What are the names of Viv Windsor-Hope's children on Emmerdale?
4. What was the name of the Italian Restaurant run by the Di Marco family on Eastenders?
5. How did Drew die in Neighbours?
6. Which Dynasty character believed that they had been abducted by aliens? Alexis, Jeff or Fallon?
7. What religion was Coronation Street's Ivy Tilsley?
8. In Dallas, what addiction haunted Sue-Ellen?
9. On Home and Away, why did Sophie run away?
10. What shop did Mick Johnson run on The Parade in Brookside?

1. Beverly Hills 90210
2. Worked the steam press in the prison laundry
3. Scott and Donna
4. Guiseppe's
5. In a riding accident
6. Fallon
7. Roman Catholic
8. Alcoholism
9. Because she was pregnant
10. A Pizza Parlour

ROUND 6
SOAP OPERAS 4

1. What was the occupation of Pippa's second husband Michael on Home and Away? A Marine Engineer, a chef or a doctor?
2. On Eastenders, who is Vicky Watt's mother?
3. Patrick Duffy played which Ewing brother on Dallas?
4. What is the name of the supermarket where Curly and Reg worked on Coronation Street?
5. What was the name of Anne Wilkinson's first love and on/off boyfriend on Neighbours?
6. Who did Tina Dingle jilt at the alter in 1995?
7. In The Archer's what was the name of the fawn that was hand-reared by Caroline Bone after its mother was killed by poachers?
8. On Beverly Hills 90210, who didn't Dylan date? Kelly, Donna or Brenda?
9. Who exchanged the first lesbian kiss on Brookside?
10. What soap briefly returned to ITV in 2001 after a thirteen year absence?

1. Blake Carrington.
2. The Woolpack
3. Simon 'Wicksy' Wicks
4. True
5. Grange Farm
6. Summer Bay
7. Aaron Spelling
8. Sinbad
9. Wentworth Detention Centre
10. Christopher

SOAP OPERAS 5

1. Which Dynasty character was originally played by Pamela-Sue Martin and then Emma Samms??
2. In Eastenders, what was the name of the wine bar that Dirty Den burnt down?
3. What did Coronation Street's Tyrone Dobbs give Jack Duckworth on his 60th birthday?
4. Which of these have NOT appeared on Brookside? Tracey Ullman, Graham Norton or Morrissey?
5. What show was a spin-off from Dallas? Knots Landing, Falcon Crest or Dynasty?
6. What is the catchphrase that is used on the annual trip to Tijuana in The O.C?
7. In Dawson's Creek, who were the first two characters to appear on the show?
8. Where is Hollyoaks set?
9. What was the name of Sharon's band that she sang in as a teenager on Eastenders?
10. What was Jane Harris's nickname on Neighbours?

1. Fallon
2. The Dagmar
3. A stuffed and mounted racing pigeon.
4. Tracey Ullman
5. Knots Landing
6. "What happens in Mexico stays in Mexico".
7. Joey and Dawson
8. Chester
9. The Banned
10. Plain Jane Super Brain

110   FILM & TV QUIZ BOOK

QUESTIONS

1. What is the name of Pauline's dog on Eastenders?
2. What street do the Neighbours live on?
3. In Dallas, Jenna Wade jilted which Ewing brother twice?
4. Coronation Street's Nicky Tilsley married which Battersby?
5. Alexis and Dex Dexter form which company in Dynasty?
6. What is the name of Lisa and Zak Dingle's daughter on Emmerdale?
7. What was Tracy Corkhill's profession on Brookside?
8. Who had a surrogate baby for Sally on Home and Away?
9. In Beverley Hills 90210 who dated both Brandon and Dylan? Donna, Kelly or Andrea?
10. Oscar-winning actor Ben Kingsley appeared in which UK soap?

ANSWERS

1. Betty
2. Ramsay Street
3. Bobby
4. Leanne
5. Lex-Dex
6. Belle
7. Hairdresser
8. Leah
9. Kelly
10. He played Ron Jenkins in Coronation Street

1. On The Archers, which Grundy tried to become a Country & Western star?
2. What was the name of Blake and Alexis's gay son on Dynasty?
3. What sort of film did Trisha star in on Emmerdale?
4. Bouncer the golden Labrador appeared on which Australian soap?
5. Who was responsible for the death of Sue and Danny Sullivan in Brookside?
6. Where do Dot and Pauline work on Eastenders?
7. What is the name of Vera's husband on Coronation Street?
8. Which were the feuding families on Dallas?
9. Where is Ryan Atwood originally from on the O.C?
10. On Hollyoaks, which character did Elize du Toit play? Lara, Izzy or Carol?

1. Eddie
2. Stephen
3. Bollywood movie
4. Neighbours
5. Barry Grant
6. The launderette
7. Jack Duckworth
8. The Ewings and the Barnes
9. Chino, California
10. Izzy

ROUND 6
SOAP OPERAS 8

1. What instrument did Sonia play on Eastenders?
2. What was the name of Steve McDonald's first wife on Coronation Street?
3. In Dynasty, Amanda Carrington married a prince from which fictional country?
4. What were the christian names of the Alessi twins on Neighbours?
5. What is the name of Katie Holmes' character on Dawson's Creek?
6. What is the name of Lindsay Corkhill's daughter on Brookside?
7. What relation was Valerie to Brenda on Beverly Hills 90210?
8. What is the name of the village in which the Archers is set?
9. Which soap did singer, Ricky Martin get his big break on?
10. What was the name of the medical practice in Doctors that burned down?

1. The trumpet
2. Vicky Arden
3. Moldavia
4. Caroline and Christina
5. Josephine 'Joey' Potter
6. Kylie
7. Cousin
8. Ambridge
9. General Hospital
10. Riverside & Best

QUESTION

1. Who was driving the car that killed Tiffany Mitchell on Eastenders?
2. What is the name of the Archers' theme music? Greensleeves, Barwick Green or The Green Man?
3. Which Oscar-nominated actress once played a pair of look-alike half sisters on As The World Turns?
4. When Lynn Scully first arrived in Neighbours, what was her occupation?
5. In Coronation Street, Deirdre Rashid was sentenced to prison for committing which crime?
6. On Dallas, who was JR Ewing's arch rival and Pam Ewing's brother?
7. What was the name of Pacey's sailboat on Dawson's Creek?
8. What was the name of Max Farnham's first wife on Brookside?
9. Which state did the Walsh family move from before arriving in Beverly Hills 90210?
10. Which Spanish-set BBC soap started in July 1992 and ended July 1993?

ANSWERS

1. Frank Butcher
2. Barwick Green
3. Julianne Moore
4. Hairdresser
5. Fraud
6. Cliff Barnes
7. True Love
8. Susannah
9. Minnesota
10. Eldorado

QUESTIONS

1. What is the name of Seth Cohen's boat on The O.C?
2. What nickname did Neighbours' Lou Carpenter give Harold Bishop in school?
3. Which actor links The Boys From The Black Stuff, The Royale Family and Brookside?
4. Who performs the opening song, 'I Don't Want To Wait' on Dawson's Creek?
5. In Eastenders, which football team scouted the talents of Aidan Brosnan, played by Sean Maguire?
6. What was the name of the diner where Lucy Ewing worked in Dallas?
7. In Coronation Street, what did Terry Duckworth sell to his in-laws on Christmas Eve 1993?
8. In which Aussie soap did actress Emily Symons appear before playing Louise in Emmerdale?
9. What job did Krystle do before marrying Blake Carrington in Dynasty?
10. Which weekend ITV soap, set on a street market in Manchester, was cancelled after just 100 episodes in 1986?

ANSWERS

5. Walford Town FC
4. Paula Cole
3. Ricky Tomlinson
2. Jelly Belly
1. Summer Breeze

10. Albion Market
9. She was his secretary
8. Home & Away
7. His son
6. The Hot Biscuit

1. What was the name of the diner in Beverly Hills 90210?
2. Where did Dawson and his friends go to school in Dawson's Creek
3. What instrument did Harold Bishop play in Neighbours?
4. Which two actresses played the character of 'Miss Ellie' in Dallas?
5. What disaster befell the community of Emmerdale Farm in December 1993, killing off four of its characters?
6. Which characters punctured a waterbed whilst in the throes of passion in Coronation Street?
7. What was the name of the minicab firm that Sue and Ali Osman ran from their café in Eastenders?
8. How did Adam try to kill his brother Jeff in Dynasty?
9. Who was Bobby Simpson's real father in Home and Away?
10. In which soap did Elaine Paige and Malcolm McDowell appear?

1. The Peach Pit
2. Capeside High
3. The tuba
4. Barbara Bel Geddes and Donna Reed
5. A plane crashed into the village
6. Reg Holdsworth and Maureen Naylor
7. Ozcabs
8. By having his office painted with poisonous paint
9. Her arch enemy Donald Fisher
10. Crossroads

QUESTIONS

1. In Dawson's Creek what was the name of Jen's friend who drowned?
2. Which pop star played Beth in "Neighbours" between 1991 and 1993?
3. What is the name of the fictional tube station in "Eastenders"?
4. What relation were Brenda and Brandon on Beverly Hills 90210?
5. Originally starting as a six part series in 1957, "Calling Nurse Roberts", what was Britain's first serial to be shown twice-weekly?
6. In which year did "Emmerdale Farm" become "Emmerdale"? 1989, 1992 or1995?
7. First shown on the station's launch night, what is Channel 5's first original soap opera called?
8. Where in Orange County is The O.C set?
9. Which soap opera is set in the fictional county of Borsetshire?
10. What was the name of Jimmy Corkhill's dog on Brookside?

ANSWERS

1. Abby
2. Natalie Imbruglia
3. Walford East
4. Brother and sister
5. Emergency – Ward 10
6. 1989
7. Family Affairs
8. Newport Beach
9. The Archers
10. Cracker

1. What was Charlie Cotton's profession in Eastenders?
2. In Brookside, what was the name of Sammy Daniels's husband?
3. Neighbours was set in which town?
4. What was the name of Gina Patrick's girlfriend in Hollyoaks?
5. Pacey slept with one of his teachers on Dawson's Creek, what subject did she teach?
6. Which soap was set in the fictional Scottish village of Glendarroch
7. Brad Pitt appeared in which US soap as 'Randy'? Dallas, Dynasty or Beverley Hills 90210?
8. Guy Pearce starred in which two soaps before finding movie fame?
9. Which Tony-winning Broadway actress played the part of Dominique Devereaux in Dynasty?
10. In 1982, who did Emmerdale's Jack Sugden marry? Pat Merrick, Kathy Merrick or Kathy Bates?

1. Long-distance lorry driver
2. Owen
3. Erinsborough
4. Emily
5. English
6. Take the High Road (later known as High Road).
7. Dallas, playing Charlie's boy-friend
8. Neighbours as Mike Young and Home and Away as David Croft
9. Diahann Carrol
10. Pat Merrick

QUESTIONS

1. What was the name of Brookside's first spin-off 'soap bubble'?
2. What was the name of the magazine that Neighbours Lucy Robinson posed naked for?
3. What was the name of the hostess at the Oil Baron's Club in Dallas?
4. What was Jambo's car in Hollyoaks was shaped like?
5. What is the name of Nigel Pergetter's Haris Hawk in The Archers?
6. Dannii Minogue starred as Emma Jackson in which Aussie soap?
7. In the series finale of Dawson's Creek, what is the name of Jack's boyfriend?
8. Who did Butch Dingle marry on his hospital deathbed in Emmerdale?
9. Which male character on Dawson's Creek entered the Miss Capeside pagent?
10. What was Coronation Street's Judy Mallet doing when she died?

ANSWERS

1. Damon and Debbie
2. Ambrosia
3. Dora-Mae
4. A mouse
5. Eleanor
6. Home and Away
7. Doug
8. Emily Wylie
9. Pacey
10. Hanging out the washing in the back yard

QUESTIONS

1. In Hollyoaks, what was the name of the Big Brother style reality show that Adam filmed as his media project?
2. What was the name of Lauren Carpenter's horse in neighbours? Chuck A Mental, Throw A Wobbler or Lose the Plot?
3. What was the name of Pam and Cliff Barnes's father in Dallas?
4. For which crime did Emmerdale's Sam Dingle get sentenced to 4 months imprisonment?
5. In Dawson's Creek, what does Dawson want to be when he graduates?
6. On which street is Family Affairs set?
7. On The O.C what is the name of Seth's grandfather?
8. Which university did Ken Barlow attend in Coronation Street?
9. What was the name of the night club opened by Barry Grant in Brookside?
10. How many children did Eastenders' matriarch Lou Beale have?

ANSWERS

1. The Fish Tank
2. Chuck A Mental
3. Willard 'Digger' Barnes
4. Hit and run
5. A film director
6. Stanley Street
7. Caleb Nichol
8. Manchester, where he gained a second class degree in History and English.
9. La Luz
10. Five. Pauline, Peter, Kenny, Ronnie and her illegitimate daughter Maggie

QUESTIONS

1. What was the name of Cliff Barnes's mother's company in Dallas?
2. Who was Cecil Colby having sex with when he had a heart attack in Dynasty?
3. What is the name of the hotel in Neighbours?
4. Where did Ricky and Sam Butcher get married?
5. In which soap did 'The Freak' and 'Vinegar Tits' appear?
6. Which soap links Ben Kingsley, Joanna Lumley and Davy Jones?
7. What was the name of Ron Dixon's general store on Brookside?
8. Whose family own The Icehouse in Dawson's Creek? Joey's Pacey's or Dawson's?
9. In Hollyoaks, whose mother did Finn marry?
10. In the Archers, what is the name of Tony's vintage car?

ANSWERS

1. Wentworth Tool & Die
2. Alexis.
3. Lassiter's
4. Gretna Green
5. Prisoner Cell Block H
6. Coronation Street
7. The Trading Post
8. Joey's.
9. Tony Hutchinson
10. Esme.

ULTIMATE
BRAINBUSTER

ROUND 7
COPS &

ROBBERS

QUESTIONS

1. In which cop TV show did Johnny Depp start his career?
2. What actor played Perry Mason and Ironside?
3. What gangster film stars Christopher Walken and Benicio Del Toro?
4. Who was the shaven headed lollipop sucking cop?
5. Who plays the title role in Johnny Dangerously?
6. Nicolas Cage plays the role Richard Widmark played in the remake of which film?
7. Who plays Detective John Kelly in NYPD Blue?
8. How many spin-offs does CSI have and in which cities?
9. What was Natalie Portman's film debut?
10. What UK TV series ran from 1977 till 1983 and starred Lewis Collins as William Bodie?

ANSWERS

1. 21 Jump Street
2. Raymond Burr
3. The Funeral
4. Kojak
5. Michael Keaton
6. Kiss of Death
7. David Caruso
8. Two: Miami and New York
9. Leon: The Professional
10. The Professionals

ROUND 7
COPS & ROBBERS 2

1. Who played Starsky in the film Starsky & Hutch?
2. Who played DI Jack Regan in The Sweeny?
3. Softly, Softly was a spin off from what TV series?
4. In Double Indemnity, Fred MacMurray is talked into murder by whom?
5. Who sang the theme song to the film Shaft?
6. In which Dirty Harry sequel does Jim Carrey play a small role?
7. What pop star played a Train Robber in 1988?
8. In which John Boorman film does Brendan Gleason play a thief?
9. In which film does Jude Law play Jude, Sadie Frost play Sadie, and Ray Winstone plays Ray?
10. In Ordinary Decent Criminal, which actor plays an Irish gangster?

1. Ben Stiller
2. John Thaw
3. Z Cars
4. Barbra Stanwyck
5. Isaac Hayes
6. Dead Pool
7. Phil Collins
8. The General
9. Love Honour Obey
10. Kevin Spacey

ROUND 7
COPS & ROBBERS 3

QUESTIONS

1. What is the name of James Caan's character in The Godfather?
2. Who directed Robocop?
3. Who is Kaiser Soze?
4. In which film does Al Pacino shout 'Attica!'?
5. In which film does Sidney Poitier say 'they call me Mr. Tibbs'?
6. What caper film directed by Stanley Kubrick stars Sterling Hayden?
7. In which film does Clint Eastwood carefully plan his escape from an Island prison?
8. Russell Crowe and Guy Pearce star in which Los Angeles based crime film?
9. Orson Welles directed Charlton Heston in which 1958 crime classic?
10. What links Cannon and Rocky and Bulwinkle?

ANSWERS

1. Sonny Corleone
2. Paul Verhoven
3. Kevin Spacey
4. Dog Day Afternoon
5. In The Heat of the Night
6. The Killing
7. Escape from Alcatraz
8. L.A. Confidential
9. Touch of Evil
10. Actor William Conrad played Cannon and was the voice of Bulwinkle

ROUND 7
COPS & ROBBERS 4

1. What TV series stars Dennis Farina as Lt Michael Turello?
2. What was Dixon of Dock Green's catchphrase?
3. What TV series has Sgt. Joe Friday?
4. What film starred both James Cagney and Humphrey Bogart?
5. Who plays the family consigliore in Godfather 1 & 2?
6. What is Al Pacino's character's profession in Heat?
7. Where does Michael go to hide after killing Sollozo in Godfather 1?
8. Jackie Brown is based on a novel by which acclaimed crime novelist?
9. In Get Shorty, which industry does John Travolta want to enter?
10. Sidney Sheldon created which TV series starring Robert Wagner?

Crime Story
2. Evening all.
3. Dragnet
4. The Roaring Twenties
5. Robert Duvall
6. Detective
7. Sicily
8. Elmore Leonard
9. Film
10. Hart to Hart

ROUND 7
COPS & ROBBERS 5

QUESTIONS

1. Who plays Michael Corleone's daughter in Godfather Part 3?
2. What was the longest running cop series in the history of American TV?
3. What was the second film to star Paul Newman and Robert Redford?
4. What actor plays Detective Columbo?
5. What TV producer created Hill Street Blues and L.A. Law?
6. Who starred in the classic crime drama The Big Sleep?
7. What studio was responsible for most of the gangster films of the 1930s and 40s?
8. Who composed the music for The Godfather?
9. What was the first ever crime story told on film?
10. Since what year has The Bill been running?

ANSWERS

1. Sofia Coppola
2. Hawaii Five-0
3. The Sting
4. Peter Falk
5. Steven Bocho

6. Humphrey Bogart
7. Warner Brothers
8. Nino Rota
9. The Great Train Robbery (1903)
10. 1984.

ROUND 7
COPS & ROBBERS 6

QUESTIONS

1. What was the name of Don Johnson's character in Miami Vice?
2. What actress stars as Bob Hoskins' wife in The Long Good Friday?
3. What rock star has a role in Lock, Stock and Two Smoking Barrels?
4. In which TV series does Robbie Coltrane play a psychologist called Fitz?
5. Goodfellas was based on the true story of which mobster?
6. In which series of films does Clint Eastwood play Detective Harry Callahan?
7. Who plays forensic detective Grissom on CSI?
8. Gene Hackman plays Popeye Doyle in what 70s crime classic?
9. In which film does Al Pacino play retired gangster Carlito Brigante?
10. 'Say Hello to my little friends' is a line from which gangster film?

ANSWERS

1. Detective James 'Sonny' Crocket
2. Helen Mirren
3. Sting
4. Cracker
5. Henry Hill
6. Dirty Harry
7. William Petersen
8. The French Connection
9. Carlito's Way
10. Scarface

1. Who plays Captain McCluskey in The Godfather?
2. Who is Michael Douglas' partner in Black Rain?
3. What crime film was the Coen brothers debut feature?
4. In which film does Al Pacino shout 'Attica!'?
5. What actor played Baretta?
6. In which TV show do two attractive cops ride motorcycles?
7. Patrick McGoohan won two Emmys as a guest star on which mystery show?
8. What actor played the title role in the film Donnie Brasco?
9. 'Just the facts Ma'am' is which Detective's catchphrase?
10. What TV producer created NYPD Blue, Hill Street Blues and L.A. Law?

1. Asphalt Jungle
2. Andy Garcia
3. Blood Simple
4. Dog Day Afternoon
5. Robert Blake
6. Chips
7. Columbo
8. Johnny Depp
9. Joe Friday – Dragnet
10. Steven Bocho

ROUND 7
COPS & ROBBERS 8

1. Marilyn Monroe has a small part in a classic heist film from 1950 – which one?
2. Who starred in Angels with Dirty Faces?
3. What writer created the Sherlock Homes character?
4. In which film does Humphrey Bogart play Detective Philip Marlowe?
5. In which film does Clint Eastwood play a western cop in the big city?
6. Bruce Willis is taking care of a child with special powers in which film?
7. The Self-Preservation Society is a song from what film?
8. What is Pam Grier's profession in the film Jackie Brown?
9. In which film do Jada Pinkett and Queen Latifah rob banks?
10. In Blood Work, where does Clint Eastwood live?

1. Asphalt Jungle
2. James Cagney
3. Sir Arthur Conan Doyle
4. The Big Sleep
5. Coogan's Bluff
6. Mercury Rising
7. The Italian Job
8. Flight attendant
9. Set it Off
10. A boat

1. In the film G Men, who does James Cagney play?
2. In what Morgan Freeman film does he go looking for a congressman's daughter?
3. What platinum blonde plays Cagney's girlfriend in The Public Enemy?
4. Alan Ladd starred with which actress in a series of film noir movies in the 1940s?
5. What was Woody Allen's only film to deal with gangsters?
6. Warren Beatty played on film a famous comic strip detective in 1990- what film was it?
7. The Big Sleep was remade in 1977 with which actor as Philip Marlowe?
8. What American TV show divides half the show in half: detective work and legal prosecution?
9. Which ex-NYPD Blue actor plays the lead in CSI: Miami?
10. In which film does Johnny Depp go undercover with the New York mob?

1. An FBI agent
2. Along Came a Spider
3. Jean Harlow
4. Veronica Lake
5. Bullets Over Broadway
6. Dick Tracy
7. Robert Mitchum
8. Law & Order
9. David Caruso
10. Donnie Brasco

ROUND 7
COPS & ROBBERS 10

QUESTIONS

1. Drea de Matteo, who plays Adriana La Cerva in The Sopranos, went on to star in what sitcom?
2. What Australian film stars Anthony LaPaglia as a Detective cheating on his wife?
3. In Sexy Beast, Ben Kingsley comes to recruit which actor for one last job?
4. In which epic gangster film did Jennifer Connelly make her film debut?
5. In which Robert Rodriguez film does Johnny Depp play CIA Agent Sands?
6. What is the only film to star Michael Douglas and Catherine Zeta Jones?
7. In which decade did The Sweeny first appear?
8. Tommy Lee Jones plays American developer Cosmo in which Manchester-based crime film?
9. Clive Owen plays an unsuccessful novelist working at a casino in which Mike Hodges film?
10. In Gangster No 1, who plays the young Malcolm McDowell?

ANSWERS

1. Joey
2. Lantana
3. Ray Winstone
4. Once Upon a Time in America
5. Once Upon a Time in Mexico
6. Traffic
7. 1970s (75)
8. Stormy Monday
9. Croupier
10. Paul Bettany

ROUND 7
COPS & ROBBERS 11

QUESTIONS

1. Who wrote the book "The Godfather", on which the film is based?
2. What part was Tom Selleck due to play before he started Magnum P.I.?
3. In Lethal Weapon 4, which Hong Kong martial arts star joins the cast?
4. Michael Mann created a TV series in the 1980s famous for its use of fashion and music – what was it called?
5. Who starred alongside Karl Malden in Streets of San Francisco?
6. Who played Kojak?
7. Robert DeNiro plays the younger version of which character in Godfather II?
8. How many sequels of Robocop are there?
9. What does CSI stand for?
10. What links Bergerac and Groundhog Day?

ANSWERS

1. Mario Puzo
2. Indiana Jones
3. Jet Li
4. Miami Vice
5. Michael Douglas
6. Telly Savalas
7. Vito Corleone
8. Two
9. Crime Scene Investigation
10. Both featured scores by George Fenton

134 FILM & TV QUIZ BOOK

ROUND 7
COPS & ROBBERS 12

QUESTIONS

1. What is the name of the company that builds Robocop?
2. Who plays the lead gangster in Little Caesar?
3. What nationality was Al Pacino's character in Scarface?
4. What two Academy-Award winning actors starred in Heat?
5. What two pop stars played the Kray brothers in the film The Krays?
6. What year was The Godfather Part II released?
7. What James Bond has a small part in The Long Good Friday?
8. Who played the gangster in Public Enemy?
9. What was the first all-female buddy cop show?
10. Who is the writer behind Prime Suspect?

ANSWERS

1. OCP
2. Edward G. Robinson
3. Cuban
4. Robert DeNiro and Al Pacino
5. Gary and Martin Kemp
6. 1974
7. Pierce Brosnan
8. James Cagney
9. Cagney and Lacey
10. Lynda La Plante

1. In what Hawaiian city was Hawaii Five-O set?
2. What is the name of Michael's older brother who betrays him in Godfather II?
3. Apart from Heat, which other film were Robert DeNiro and Al Pacino in together?
4. What character in The Long Good Friday says: 'You don't crucify people! Not on Good Friday!'?
5. In which film does Al Pacino play a cop fighting police corruption?
6. What country does Gene Hackman travel to in French Connection II?
7. In which film does Harry Connick Jr. play a convicted killer?
8. Who stars opposite Denzel Washington in The Bone Collector?
9. Who plays both the suspect and Al Pacino's love interest in Sea of Love?
10. What actor stars in the series The Rockford Files?

1. Honolulu
2. Fredo
3. Godfather Part II
4. Harold (Bob Hoskins)
5. Serpico
6. France
7. Copycat
8. Angelina Jolie
9. Ellen Barkin
10. James Garner

ROUND 7
COPS & ROBBERS 14

1. In Snatch, which role does Brad Pitt play?
2. Who plays Mel Gibson's partner in the Lethal Weapon series?
3. The American remake of Get Carter starred which action hero of the 1980s?
4. Tarantino's first film concerned a group of diamond thieves. What was it called?
5. What 1998 crime film starring George Clooney is based on an Elmore Leonard novel?
6. In Get Carter, which city does Michael Caine travel to in order to find his brother's killer?
7. George Clooney stars in this remake of an original Rat Pack film from the 60s – what is it called?
8. In The Professionals, who was the head of CI5?
9. The Maltese Falcon is based on a book by which author?
10. The villain in Lethal Weapon 2 is the Ambassador of which country?

1. Mickey O'Neil, a gypsy boxer
2. Danny Glover
3. Sylvester Stallone
4. Reservoir Dogs
5. Out of Sight
6. Newcastle
7. Ocean's 11
8. Cowley
9. Dashiell Hammet
10. South Africa

QUESTIONS

1. What is the Law & Order spin-off that deals with sexual cases called?
2. What 1950s film has a heist that takes place at a racecourse?
3. What religious institution does Michael Corleone deal with in Godfather Part III?
4. What is the first sequel to Dirty Harry called?
5. What cop show did William Shatner star in during the 1980s?
6. In which decade did Charlie's Angels debut on American television?
7. The Defiant Ones chained Tony Curtis and which black actor together?
8. 'What we've got here…is failure to communicate' is a line from which film?
9. In which film does Clint Eastwood pair up with Charlie Sheen?
10. From which British crime film is the line 'Can everyone stop getting shot'?

ANSWERS

1. Special Victims Unit
2. The Killing
3. The Vatican
4. Magnum Force
5. T.J. Hooker
6. 1970s (76)
7. Sidney Poitier
8. Cool Hand Luke
9. The Rookie
10. Lock, Stock and Two Smoking Barrels

138   FILM & TV QUIZ BOOK

ROUND 7
COPS & ROBBERS 16

1. Joe Pesci stars as stuttering Leo Getz, a criminal witness, in what film?
2. Who plays Dillinger in the 1973 John Millius film of the same name?
3. In which film does Denzel Washington show Ethan Hawke the ropes on the streets of L.A.?
4. In Dog Day Afternoon, what does Al Pacino need the money from the bank raid for?
5. Steve McQueen is on the run with which former model in The Getaway?
6. What is Britain's number one TV police drama?
7. What long-running British cop show is set in Yorkshire?
8. What film stars Al Pacino as a true-story New York cop that refused to take bribes?
9. In which city does Casino take place?
10. What gun does Detective Harry Callahan use?

1. Lethal Weapon 2
2. Warren Oates
3. Training Day
4. Sex change operation for his lover
5. Ali McGraw
6. The Bill
7. Heartbeat
8. Serpico
9. Las Vegas
10. Magnum 44

1. In which sequel does Gene Hackman become addicted to heroin?

2. In which film does Al Pacino infiltrate gay clubs in New York?

3. Who plays the sexy blonde suspect in Sea of Love?

4. In which film is Denzel Washington a disabled cop?

5. What film stars Humphrey Bogart as Sam Spade?

6. This Gun for Hire was Alan Ladd and which actress' first pairing?

7. The Coen Brothers remade The Big Sleep – what did they call it?

8. Who created the character Philip Marlowe?

9. What was Sergio Leone's only gangster film?

10. For which film did Frances McDormand win an Academy Award?

10. Fargo
9. Once Upon a Time in America
8. Raymond Chandler
7. The Big Lebowski
6. Veronica Lake

5. Maltese Falcon
4. The Bone Collector
3. Ellen Barkin
2. Cruising
1. French Connection II

QUESTIONS

1. What famous pop star did the director of Lock, Stock and Two Smoking Barrels go on to marry?
2. What actor plays Tony Soprano?
3. In which film does John Travolta use Hugh Jackman to steal billions of dollars?
4. What role does Lorraine Bracco play in Goodfellas?
5. 'King Kong ain't got nothin' on me' is a line uttered by Denzel Washington in which film?
6. In which film do Dennis Hopper and Christopher Walken have a funny but ultimately deadly scene?
7. What Steve McQueen film features a car chase scene on the streets of San Francisco?
8. In which film did Peter Lorre and Humphrey Bogart first meet?
9. In which city does the film Heat take place?
10. What Academy Award nominated film is based on a James Ellroy novel?

ANSWERS

1. Madonna
2. James Gandolfini
3. Swordfish
4. Henry Hill's wife Karen
5. Training Day
6. True Romance
7. Bullitt
8. Maltese Falcon
9. Los Angeles
10. L.A. Confidential

ULTIMATE BRAINBUSTER

ROUND 8

ROMANCE

1. What is the highest grossing film of all time?
2. Celine Dion sang the theme song to Titanic. What was its name?
3. What famous phrase did James Cameron say upon receiving the Oscar for Titanic?
4. What is Kate Winslet's first name in Titanic?
5. What actor plays Romeo in Baz Luhrman's Romeo & Juliet?
6. What 2000 romantic musical was nominated for Best Picture but not Best Director?
7. Nicole Kidman recorded a duet with which British pop star?
8. What director better known for gangster films directed The Age of Innocence?
9. Daniel Day Lewis falls for Madeleine Stowe in which epic romantic drama?
10. Somersby is a remake of what French film?

1. Titanic
2. My Heart will Go On
3. I'm the King of the World!
4. Rose
5. Leonardo DiCaprio
6. Moulin Rouge
7. Robbie Williams
8. Martin Scorsese
9. The Last of the Mohicans
10. Return of Martin Guerre

ROUND 8
ROMANCE 2

1. Where does Brief Encounter mainly take place?
2. What is Julia Roberts' profession in Pretty Woman?
3. In which film does Richard Gere carry Debra Winger in his arms in the final scene?
4. In which film does Richard Gere play a male prostitute?
5. Rhett Butler is played by which actor in Gone with the Wind?
6. 'Love means never…" complete the quote from Love Story
7. Who played the title role in the film Marty?
8. Ralph Fiennes' brother Joseph plays Shakespeare in which film?
9. What was Anthony Minghella's first film?
10. In which film does Ava Gardner play a statue that comes to life?

1. Train station
2. Prostitute
3. An Officer and a Gentleman
4. American Gigolo
5. Clark Gable
6. '…having to say you're sorry'
7. Ernest Borgnine
8. Shakespeare in Love
9. Truly Madly Deeply
10. One Touch of Venus

1. What university does Oliver attend in Love Story?
2. How many Oscars did romantic epic Gone with The Wind win?
3. The English Patient is set during which war?
4. Opposite which English actor did Marilyn star in The Prince and The Showgirl?
5. What film features a romantic scene at a pottery wheel?
6. In which film do Catherine Zeta Jones and Julia Roberts play sisters?
7. Who plays Marianne Dashwood in Sense and Sensibility?
8. In which film does John Cusack hold a boom box in the air playing Peter Gabriel?
9. 'You complete me' is a line from which Tom Cruise film?
10. In which American city does Singles take place?

1. Harvard
2. Nine
3. World War II
4. Lawrence Olivier
5. Ghost
6. America's Sweethearts
7. Kate Winslet
8. Say Anything
9. Jerry Maguire
10. Seattle

ROUND 8
ROMANCE 4

1. What famous pop singer stars in Waiting to Exhale?
2. In which film does Audrey Hepburn play a Princess?
3. 'We'll always have Paris' is a line from which film?
4. What was Robert Redford and Barbra Streisand's only film together?
5. 'I Like to Live in America' is a song from which romantic musical?
6. Doctor Zhivago takes place during what revolution?
7. Who plays the title role in Annie Hall?
8. Who plays Henry Higgins in My Fair Lady?
9. Out of Africa won Best Picture in which year?
10. Humphrey Bogart won his only Oscar for his performance in which film opposite Katherine Hepburn?

1. Whitney Houston
2. Roman Holiday
3. Casablanca
4. The Way We Were
5. West Side Story
6. Russian Revolution
7. Diane Keaton
8. Rex Harrison
9. 1985
10. The African Queen

QUESTIONS

1. Who played Heathcliff in the 1939 version of Wuthering Heights?
2. Who plays Gene Kelly's girlfriend in Singing in the Rain?
3. In which film does Nicolas Cage play a passionate baker?
4. In which film does James Stewart suffer from fear of heights?
5. Whoopi Goldberg won an Oscar for her performance in which romantic film?
6. Who plays rebel Private Prewitt in From Here to Eternity?
7. What is Richard Gere's profession in Pretty Woman?
8. What was Henry Fonda's last film?
9. Who is the star of Now, Voyager?
10. Fay Wray was (literally) in which giant ape's hands?

ANSWERS

1. Laurence Olivier
2. Debbie Reynolds
3. Moonstruck
4. Vertigo
5. Ghost
6. Montgomery Cliff
7. Corporate Raider (buys and sells companies)
8. On Golden Pond
9. Bette Davis
10. King Kong

ROUND 8
ROMANCE 6

1. Who plays Meg Ryan's best friend in When Harry Met Sally?
2. In which film do Barbra Stanwyck and Henry Fonda meet on a ship?
3. 'The Hills Are Alive' is the opening song of which film?
4. What was The Shop Around the Corner remade as in 1998?
5. In which film does Fred Astaire have to raise $25000 to marry?
6. The King and I was remade in 1999 with which actress in the Deborah Kerr role?
7. In which film does Humphrey Bogart play Bette Davis' Irish stableman?
8. What animated Disney love story was nominated for a Best Picture Oscar in 1991?
9. Who plays Molly Brown in Titanic?
10. In which film does Clark Gable play a journalist on a road with an heiress?

1. Carrie Fisher
2. The Lady Eve
3. The Sound of Music
4. You've Got Mail
5. Swing Time
6. Jodie Foster
7. Dark Victory
8. Beauty and the Beast
9. Kathy Bates
10. It Happened One Night

QUESTIONS

1. What film tells the story of Fanny Brice?
2. Who played Anna Karenina in the first film version of the novel?
3. Who plays self-destructive actor Norman Maine in the A Star Is Born?
4. In which film does Katherine Hepburn play Tracy Samantha Lord?
5. Who plays Meg Ryan's friend in Sleepless in Seattle?
6. In which Hitchcock film does Cary Grant play ex-thief John Robie?
7. In The Postman Always Rings Twice who plays the Jack Nicholson part from the remake?
8. Judi Dench plays Queen Elizabeth in what film?
9. In Bringing Up Baby, what is Baby?
10. What film features the love triangle of Montgomery Clift, Elizabeth Taylor and Shelly Winters?

ANSWERS

1. Funny Girl!
2. Greta Garbo
3. James Mason
4. Philadelphia Story
5. Rosie O'Donnell
6. To Catch a Thief
7. John Garfield
8. Shakespeare in Love
9. A Leopard
10. A Place in the Sun

ROUND 8
ROMANCE 8

QUESTIONS

1. Who played the Humphrey Bogart part in the remake of Sabrina?
2. The Warren Beatty film Reds tells the story of which American journalist?
3. What continent do Audrey Hepburn and Albert Finney travel through in Two for the Road?
4. Who was the unexpected guest in Guess Who's Coming to Dinner?
5. Who was the star of Picnic?
6. To Have and Have Not takes place in which Island?
7. Breakfast at Tiffany's is based on a novel by which author?
8. In which film does Jack Lemmon fall for elevator operator Shirley Mclaine?
9. What black and white Woody Allen film stars Mariel Hemingway as his teenage love interest?
10. In which film does Marlon Brando scream 'Stella!'?

ANSWERS

1. Harrison Ford
2. John Reed
3. Europe
4. Sidney Poitier
5. William Holden
6. Martinique
7. Truman Capote
8. The Apartment
9. Manhattan
10. Streetcar Named Desire

QUESTIONS

1. In which film does Ryan O'Neal play Professor Howard Bannister?
2. Emma Thompson won an Oscar in what category for Sense and Sensibility?
3. What is the name of Steve Martin's version of Cyrano de Bergerac?
4. What 1942 film was the first pairing of Spencer Tracy and Katherine Hpeburn?
5. What actor plays the title role in The American President?
6. Irene Dunne and Cary Grant star in which classic screwball comedy as two divorcees?
7. In which film does Jon Voight return from the Vietnam war in a wheelchair to Jane Fonda?
8. The Goodbye Girl and Barefoot in the Park are both films based on plays by what playwright?
9. Who is Harrison Ford's Amish love interest in Witness?
10. William Holden is in which Chinese city in the film Love is a Many-Splendoured Thing?

ANSWERS

1. What's Up Doc?
2. Best Adapted Screenplay
3. Roxanne
4. Woman of the Year
5. Michael Douglas
6. The Awful Truth
7. Coming Home
8. Neil Simon
9. Kelly McGillis
10. Hong Kong

ROUND 8
ROMANCE 10

1. What song does Sam play for Humphrey Bogart in Casablanca?
2. In Casablanca, what will Ilsa and Rick always have?
3. What is the name of the club that Humphrey Bogart owns in Casablanca?
4. In Casablanca, Humphrey ends up with whom?
5. On which film did Humphrey Bogart and Lauren Bacall meet?
6. Who played the lead in The English Patient?
7. What is Ralph Fiennes' profession in The English Patient?
8. In which year did The English Patient win Best Picture Oscar?
9. Shirley McLaine and Jack Lemmon star in this romantic comedy set in New York City – what is it called?
10. Gregory Peck stars opposite whom in Roman Holiday?

1. As Time Goes By
2. Paris
3. Rick's Place
4. Captain Louis Renault
5. To Have and Have Not
6. Ralph Fiennes
7. Mapmaker
8. 1997
9. The Apartment
10. Audrey Hepburn

1. Who directed Brief Encounter?
2. How much does Richard Gere pay Julia Roberts for a week's company in Pretty Woman?
3. What disease afflicts towards the end of Love Story?
4. What actor won a Best Supporting Oscar for An Officer and a Gentleman?
5. Before Colin Firth, which legendary British actor played Darcy in Pride and Prejudice?
6. Who plays groupie Penny Lane in Almost Famous?
7. Who does Carrie Bradshaw end up with at the end of Sex & The City?
8. Where do George Clooney and Jennifer Lopez meet in Out of Sight?
9. 'You make me want to be a better man' is a Jack Nicholson-delivered line from which film?
10. What actress takes care of Ralph Fiennes in The English Patient?

1. David Lean
2. $3000
3. Cancer
4. Louis Gossett Junior
5. Lawrence Olivier
6. Kate Hudson
7. Big (John)
8. In the trunk of a car
9. As Good As It Gets
10. Juliette Binoche

ROUND 8
ROMANCE 12

1. What day of the week do the lovers meet regularly in Brief Encounter?
2. What film reunited the stars and director of Pretty Woman?
3. Who was the male lead of Love Story?
4. Gone with the Wind is set during what war?
5. In which city does Before Sunset take place?
6. In which Merchant-Ivory film do Emma Thompson and Anthony Hopkins have a platonic affair?
7. The Piano features a sordid love affair between Holly Hunter and whom?
8. In which film do James Spader and Susan Sarandon have an intense relationship?
9. The Unbearable Lightness of Being takes place in Czechoslovakia in what year?
10. Who plays Daniel Day Lewis' love interest in The Last of the Mohicans?

1. Thursdays
2. Runaway Bride
3. Ryan O'Neal
4. Civil War
5. Paris
6. The Remains of the Day
7. Harvey Keitel
8. White Palace
9. 1968
10. Madeleine Stowe

1. What is the first name of Ali McGraw's character in Love Story?
2. What part of the military is Richard Gere's character enrolled in An Officer and a Gentleman?
3. Dangerous Liaisons was remade as a teenage film with Sarah Michelle Gellar – what was it called?
4. Helena Bonham Carter must choose between Daniel Day Lewis and Julian Sands in which film?
5. Who plays a mermaid in the romantic comedy Splash?
6. Sigourney Weaver and Mel Gibson have an affair amidst political turmoil in which film?
7. Who said: 'I lurve you, loave you, I luff you'?
8. In which film do Ryan O'Neal and Barbra Streisand mock Love Story in the final scene?
9. Who directed the hugely successful 1968 version of Romeo & Juliet?
10. Who plays the title role in Doctor Zhivago?

1. Jennifer
2. Navy
3. Cruel Intentions
4. A Room With a View
5. Daryl Hannah
6. The Year of Living Dangerously
7. Woody Allen in Annie Hall
8. What's Up Doc?
9. Franco Zeffereli
10. Omar Sharif

ROUND 8
ROMANCE 14

1. Who stars opposite Elizabeth Taylor in Cat on A Hot Tin Roof?
2. From Here to Eternity was based in a popular novel by which author?
3. What was James Dean's last film?
4. Who starred in Peyton Place?
5. At the top of which building do Cary Grant and Deborah Kerr meet in An Affair to Remember?
6. Splendor in the Grass had Natalie Wood and which young star as lovers?
7. In which film does Bette Midler play CC Bloom?
8. In which film does Julia Roberts take care of terminally-ill Campbell Scott?
9. In the film Shadowlands, which writer does Anthony Hopkins play?
10. In The Bridges of Madison County, which country was Meryl Streep's character from?

1. Paul Newman
2. James Jones
3. Giant
4. Lana Turner
5. Empire State Building
6. Warren Beatty
7. Beaches
8. Dying Young
9. C.S. Lewis
10. Italy

1. Who does Bridget Jones end up with in the end?
2. In While You Were Sleeping, Sandra Bullock saves Bill Pullman from what?
3. What American actress does Hugh Grant fall for in Four Weddings and a Funeral?
4. Ralph Fiennes falls for which actress in the film Maid in Manhattan?
5. Kate Beckinsale falls for what actor in the film Serendipity?
6. In which year is the film The Wedding Singer set?
7. In Love Actually, which role does Hugh Grant play?
8. What is the name of Hugh Grant's flatmate in Notting Hill?
9. In A Matter of Life and Death, what is David Niven's profession?
10. What actor plays Darcy in both Bridget Jones's Diary and Pride and Prejudice?

1. Darcy
2. Train accident
3. Andie McDowell
4. Jennifer Lopez
5. John Cusack
6. 1985
7. Prime Minister
8. Spike
9. Pilot
10. Colin Firth

ROUND 8
ROMANCE 16

1. What silent movie star was the first male romantic screen idol and known as the 'Latin Lover'?
2. Who was Greta Garbo's on and off screen lover for many years?
3. What male star was Joan Crawford's romantic partner on screen for eight films?
4. Jean Harlow starred in which Howard Hughes production as the love interest of two ace pilots?
5. Who played opposite Clark Gable in the film It Happened One Night?
6. His Girl Friday was remade in 1974 with what name?
7. William Powell plays a homeless man hired as a butler in what classic film?
8. Who starred in the remake of Mr. Deeds Comes to Town in the Gary Cooper part?
9. In which film do Irene Dunn and Cary Grant squabble non-stop only to end up together?
10. Henry Fonda stars opposite which actress in The Lady Eve?

1. Valentino
2. John Gilbert
3. Clark Gable
4. Hell's Angels
5. Claudette Colbert
6. Front Page
7. My Man Godfrey
8. Adam Sandler
9. The Awful Truth
10. Barbra Stanwyck

1. What disease does Debra Winger have in Terms of Endearment?
2. Which daughter of Charlie Chaplin starred in Doctor Zhivago?
3. Which singer and actress starred in and directed The Prince of Tides?
4. Who falls for Meryl Streep in Bridges of Madison County?
5. Ryan O'Neal's affair with Ali McGraw started on the set of which film?
6. Ever After was a remake of which fairy story?
7. Who does Jim Carrey fall in love with in Eternal Sunshine of the Spotless Mind?
8. What character does Brad Pitt play in Meet Joe Black?
9. Barbra Streisand stars in The Way We Were with which actor?
10. What film stars Daniel Day Lewis and Heather Bonham Carter?

ROUND 8
ROMANCE 18

1. What author does Meryl Streep play in Out of Africa?
2. What film chronicles the romance between Sid Vicious and girlfriend Nancy?
3. In which film do Gene Tierney and Rex Harrison fall impossibly in love?
4. What character does Cary Grant play in The Bishop's Wife?
5. What actress plays Emma in the Jane Austen adaptation from 1995?
6. What actress wrote Sense and Sensibility?
7. What film stars Brad Pitt and Julia Roberts as a squarreling couple?
8. In which film do John Cusack and Catherine Zeta Jones play an on screen and off screen lovers?
9. In which Spike Lee film does interracial love between Annabella Sciora and Wesley Snipes take place?
10. Who plays Kevin Costner's father in Message in a Bottle?

1. Isak Dinesen
2. Sid & Nancy
3. The Ghost and Mrs Muir
4. An angel
5. Gwyneth Paltrow
6. Emma Thompson
7. The Mexican
8. America's Sweethearts
9. Jungle Fever
10. Paul Newman

ULTIMATE
BRAINBUSTER

ROUND 9
MUSICALS

1.   What actress played murdered pop star Selena?
2.   What is the name of the Madonna film with Antonio Banderas directed by Alan Parker?
3.   'No one puts baby in the corner' is from which film?
4.   What film chronicles the story of an Irish soul band?
5.   In which film does Kevin Bacon make a whole town dance?
6.   Who is the star of Flashdance?
7.   What is the name of the first Beatles film?
8.   Also director of The Sound of Music, who directed West Side Story?
9.   What musical stars Michael Douglas?
10.   Who wrote the musical Pennies from Heaven?

1. Jennifer Lopez
2. Evita
3. Dirty Dancing
4. The Commitments
5. Footloose
6. Jennifer Beals
7. Hard Day's Night
8. Robert Wise
9. A Chorus Line
10. Dennis Potter

ROUND 9
MUSICALS 2

1. In what decade were The Sound of Music, West Side Story and My Fair Lady made?
2. What famous 1960's model starred in Ken Russell's The Boy Friend?
3. Who plays Neil Diamond's father in the 1980 version of The Jazz Singer?
4. What was River Phoenix's last film?
5. What was the first film to use a rock n' roll song?
6. Who wrote the music for Jesus Christ Superstar?
7. What composer wrote the music for the Disney musicals Little Mermaid, Beauty and the Beast and Aladdin?
8. What Japanese film was remade with Richard Gere and Jennifer Lopez in 2004?
9. What musical stars Peter O'Toole and Sophia Loren
10. What black singer starred in the 1943 version of Stormy Weather?

10. Lena Horne
9. Man of la Mancha
8. Shall We Dance?
7. Alan Menken
6. Andrew Lloyd Webber and Tim Rice

5. Blackboard Jungle
4. A Thing Called Love
3. Laurence Olivier
2. Twiggy
1. The 1960s

1. How many films did Elvis Presley make?
2. Who starred in the 1969 musical remake of Goodbye Mr Chips?
3. Who starred in the film Mr. Holland's Opus?
4. Meryl Streep is a music teacher in a New York ghetto in which film?
5. Francis Ford Coppola's first film for a major studio was a musical – what was it called?
6. Who starred in Bye Bye Birdie?
7. What was Spike Lee's only musical film?
8. Who plays Captain Von Trapp in the Sound of Music?
9. What Latin actress plays the role of Anita in West Side Story?
10. Most of the songs in An American in Paris were written by what composers?

1. 33
2. Peter O'Toole
3. Richard Dreyfuss
4. Music of the Heart
5. Finian's Rainbow
6. Janet Leigh
7. School Daze
8. Christopher Plummer
9. Rita Moreno
10. George and Ira Gershwin

ROUND 9
MUSICALS 4

1. The Red Shoes was based on a story by which children's author?
2. Who wrote the music for the Fred Astaire-Ginger Rogers film Top Hat?
3. What Disney film starring Mickey Mouse was an all-musical, no dialogue film?
4. 'We're not in Kansas anymore' is a line from which musical?
5. What Bob Fosse film is a thinly veiled autobiography of himself?
6. What musical film won an Oscar for Catherine Zeta Jones?
7. In which musical do Nicole Kidman and Ewan McGregor do all their own singing?
8. The King Arthur legend is told in what 1967 musical?
9. What Shakespeare play is West Side Story loosely based on?
10. Who played the Judy Garland part in the 1976 remake of A Star is Born?

1. Hans Christian Andersen
2. Irving Berlin
3. Fantasia
4. Wizard of Oz
5. All That Jazz
6. Chicago
7. Moulin Rouge
8. Camelot
9. Romeo & Juliet
10. Barbra Streisand

1. Who played The Spice Girls' manager in Spice World?
2. What was the name of the all-musical episode of Buffy The Vampire Slayer?
3. What 1987 film has the hit 'The Time of My Life' on its soundtrack?
4. What musical features Albert Finney as Daddy Warbucks?
5. Who won an Oscar for his role as Master of Ceremonies in Cabaret?
6. Who sang 'Well, Did You Evah' in High Society?
7. Who stars opposite Fred Astaire in The Band Wagon?
8. Arthur Freed was a musicals producer at which Hollywood studio?
9. Who co-directed Singing in the Rain with Stanley Donen?
10. Who starred in Best Little Whorehouse in Texas with Dolly Parton?

10. Burt Reynolds
9. Gene Kelly
8. MGM
7. Cyd Charisse
6. Frank Sinatra and Bing Crosby

5. Joel Grey
4. Annie
3. Dirty Dancing
2. Once More...With Feeling
1. Richard E. Grant

ROUND 9
MUSICALS 6

1. What was Audrey Hepburn's first musical?
2. What was the only musical ever filmed in 3-D?
3. What song writing duo wrote Oklahoma! And The King and I?
4. Two musicals won Best Picture Oscar in the 1950s – one was Gigi, what was the other?
5. What was Robert De Niro's only foray into the musical genre?
6. What South Park song was nominated for an Oscar?
7. What film stars Antonio Banderas as a Cuban musician in the 1950s?
8. What was Baz Luhrman's first film?
9. In what high school was Grease set?
10. What composer and lyricist created Jesus Christ Superstar?

1. Funny Face
2. Kiss Me Kate
3. Rodgers and Hammerstein
4. An American in Paris
5. New York, New York
6. Blame Canada
7. Mambo Kings
8. Strictly Ballroom
9. Rydell High School
10. Andrew Lloyd Weber and Tim Rice

1. Who directed the musical Fiddler on the Roof?
2. What war does Claude have to go to in the musical Hair?
3. What film about ballroom dancing did Baz Luhrman direct in 1992?
4. What was the only musical Clint Eastwood ever made?
5. Who starred in Hello, Dolly!?
6. In which city does On The Town take place?
7. What was Elvis Presley's first film?
8. Who wrote Oklahoma!
9. Who starred in Easter Parade?
10. On which film set did Judy Garland and Vincente Minnelli meet?

1. Norman Jewison
2. Vietnam
3. Strictly Ballroom
4. Paint Your Wagon
5. Barbra Streisand
6. New York
7. Love Me Tender
8. Rodgers and Hammerstein
9. Judy Garland and Fred Astaire
10. Meet Me in St Louis

ROUND 9
MUSICALS 8

1. Liza Minnelli is the daughter of which musical legend?
2. In which era is Cabaret set?
3. What was the first full length 'talkie' featuring Al Jolson?
4. Who starred as Florenz Ziegfield in The Great Ziegfield?
5. What child was a musical star in the 1930's?
6. Betty Grable was the most famous pin up girl during which war?
7. Who was Fred Astaire's most famous dancing partner?
8. In which film does Bing Crosby sing White Christmas?
9. Judy Garland was paired with which child actor for seven popular musical films?
10. Who starred in Yankee Doodle Dandee?

1. Judy Garland
2. 1930s
3. The Jazz Singer
4. William Powell
5. Shirley Temple
6. World War 2
7. Ginger Rogers
8. Holiday Inn
9. Mickey Rooney
10. James Cagney

1. What was the name of the youngest Von Trapp?
2. Who played Juan Peron in the film Evita?
3. In which era was the musical Showboat set?
4. What was the name of the special Island in South Pacific?
5. Who played the original Julie La Verne in Showboat?
6. What was the name of the school in Grease?
7. Who starred in The Jazz Singer?
8. What was MGM's first musical called?
9. In which film did the song 'Singing in the Rain' first appear?
10. What rock star did Bing Crosby sing a Christmas duet with before his death?

1. Greti
2. Jonathan Pryce
3. 1880s
4. Bali Hi
5. Helen Morgan
6. Rydell High
7. Al Jolson
8. Broadway Melody
9. The Hollywood Revue of 1929
10. David Bowie

ROUND 9
MUSICALS 10

QUESTIONS

1. Who played the Mother Abbess in Sound of Music?
2. Which song did Paul Robeson make famous from Showboat?
3. What didn't Rex Harrison do in My Fair Lady?
4. What was the classic song from Oklahoma that is now a football club anthem?
5. What was the name of the character played by Catherine Zeta Jones in Chicago?
6. Who played Millie in the film Thoroughly Modern Millie?
7. What was the name of Anna's deceased husband in The King and I?
8. Where was the Phantom of the Opera set?
9. What was the name of the girls club in Grease?
10. What was the name of Annie's dog?

ANSWERS

1. Peggy Wood
2. Ol' Man River
3. Sing
4. You'll Never Walk Alone
5. Velma Kelly
6. Julie Andrews
7. Tom
8. Paris Opera House
9. Pink Ladies
10. Sandy

MUSICALS 11

1.   Who was the biggest star of synchronised swimming films in the 1940s?
2.   In which film did Fred Astaire and Gene Kelly tap dance?
3.   What musical won the Best Picture Oscar in 1951 over A Streetcar Named Desire?
4.   Who composed the music for West Side Story?
5.   What musical has received more Oscars than any other?
6.   In which film era was Singing In The Rain set?
7.   Who starred in the film A Star is Born?
8.   For which 1956 musical did Yul Brynner win the Best Actor Oscar?
9.   What is the name of Natalie Wood's character in West Side Story?
10. In which New York neighbourhood was Saturday Night Fever set?

10. Brooklyn
9. Maria
8. The King and I
7. Judy Garland and James Mason
6. Silent Era

5. Gigi
4. Leonard Bernstein
3. An American in Paris
2. Ziegfield Follies
1. Esther Williams

ROUND 9
MUSICALS 12

QUESTIONS

1. Who had summer loving in Grease?
2. On which Shakespeare play was Kiss Me Kate based?
3. Which famous song in Cabaret was written especially for Liza Minnelli?
4. Who was the lead actress in Moulin Rouge?
5. What role did Rod Steiger play in the film version of Oklahoma?
6. The Golddiggers series of musicals were created by what director/choreographer?
7. Baby Take a Bow was which child star's first starring role?
8. Oliver! Was based on a story by which author?
9. Who starred in The Court Jester?
10. Who took the role of Colonel Pickering in My Fair Lady?

ANSWERS

1. John Travolta and Olivia Newton-John
2. The Taming of the Shrew
3. Money
4. Nicole Kidman
5. Judd Fry
6. Busby Berkeley
7. Shirley Temple
8. Charles Dickens
9. Danny Kaye
10. Wilfred Hyde Whyte

1. Who washed the man right out of her hair in South Pacific?
2. What was special about Chitty Chitty Bang Bang?
3. Who was Anna Neagle's dancing partner?
4. Who composed the music for South Pacific?
5. Which character sang 'I don't know how to love him' in Jesus Christ Superstar?
6. What musical stars Spencer Tracy and Clark Gable?
7. Who stars opposite Roy Scheider in All That Jazz?
8. Who directed The Band Wagon?
9. What is the name of the Lindsay Anderson-directed musical starring Malcolm McDowell?
10. What Marx Brothers film has them placing bets on horses?

ANSWERS

1. Mitzi Gaynor
2. The car flew
3. Michael Wilding
4. Rogers and Hammerstein
5. Mary Magdalena
6. San Francisco
7. Jessica Lange
8. Vincente Minnelli
9. O Lucky Man!
10. A Day at the Races

QUESTIONS

1. Which zodiac sign was a hit song in Hair?
2. Where was the famous race scene set in My Fair Lady?
3. Who wrote the music to Hair?
4. What Irving Berlin musical stars Fred Astaire and Bing Crosby?
5. What musical features a transsexual punk rock girl from East Berlin?
6. In which film are Dan Akroyd and John Belushi on a mission from God?
7. What was Catherine Denueve's first and only musical?
8. In what film do Gene Kelly and Frank Sinatra sing 'New York, New York'?
9. What was Jean-Luc Godard's homage to American musicals?
10. What actress plays Roxie Hart in the film Chicago?

ANSWERS

1. Aquarius
2. Ascot
3. Galt MacDermot
4. Holiday Inn
5. Hedwig and the Angry Inch
6. The Blues Brothers
7. Umbrellas of Cherbourg
8. On the Town
9. A Woman is a Woman
10. Renée Zelwegger

1. Who directed Brigadoon?
2. What was the name of the Music Man?
3. What was the name of the gambler in Show Boat?
4. Who played Fiona Campbell in Brigadoon?
5. Who played Daddy Warbucks in the 1980 film version of Annie?
6. Who choreographed Brigadoon?
7. In which town did the Music Man form his brass band?
8. What is the name of the fiddler in the film Fiddler on the Roof?
9. Who played Adam, the eldest of the brothers in Seven Brides for Seven Brothers?
10. What was the novel on which The Wizard of Oz was based?

10. The Wonderful Wizard of Oz
9. Howard Keel
8. Tevye
7. River City Iowa
6. Gene Kelly

5. Albert Finney
4. Cyd Charisse
3. Gaylord Ravenal
2. Professor Harold Hill
1. Vincente Minnelli

ROUND 9
MUSICALS 16

1.	South Park The Movie spawned an Oscar-winning song. What was it called?

2.	Who plays Groucho Marx's love interest in the Marx Brothers films?

3.	'The rain in Spain…' is from what musical?

4.	Who starred in the first European talkie, Blue Angel?

5.	'Shall We Dance' is a song from what musical?

6.	What pop duo sang 'Mrs. Robinson' for the film The Graduate?

7.	'Diamonds are a Girl's Best Friend' was sung by Marilyn Monroe in which film?

8.	What film tells the fictionalized story of Eminem's life?

9.	In which film does Jennifer Beals play a welder with a passion for dancing?

10.	In what film does Elvis Presely play a convict who becomes a rock star?

1. Blame Canada
2. Margaret Dumont
3. My Fair Lady
4. Marlene Dietrich
5. The King and I
6. Simon & Garfunkel
7. Gentlemen Prefer Blondes
8. 8 Mile
9. Flashdance
10. Jailhouse Rock

ROUND 10

AlTeRnAtIvE
cInEmA

1. What documentary follows the life of two young up and coming basketball players?
2. What was the name of the trilogy of films directed by Krystof Kielowski?
3. Which Ken Loach film takes places in Northern Ireland?
4. Who directed Don't Look Now and Walkabout?
5. Peter Weir is a filmmaker from what country?
6. What does the Spanish film title Jamon, Jamon mean?
7. What film, starring Javier Bardem, tells the true story of pro-euthanasia activist and quadriplegic Ramon Sampedro?
8. Like Water for Chocolate is set in which country?
9. Which Wim Wenders film follows a Cuban music group?
10. Which Almodovar film has a female bullfighter in a coma?

1. Hoop Dreams
2. Three Colours: Blue, White, Red.
3. Hidden Agenda
4. Nicolas Roeg
5. Australia
6. Ham, Ham
7. The Sea Inside
8. Mexico
9. Buena Vista Social Club
10. Talk to Her

ROUND 10
ALTERNATIVE CINEMA 2

1. Which art-house British filmmaker made music videos for The Smiths and The Pet Shop Boys?
2. What was Peter Greenaway's first film?
3. What American city is John Waters from?
4. What Hollywood producer is the subject of the documentary The Kid Stays in the Picture?
5. What is the alternative Festival that runs alongside Sundance called?
6. What was Michael Moore's first TV show called?
7. The films of Todd Solondz take place in which American state?
8. What was Terence Malick's follow up to Badlands called?
9. What was Steven Soderbergh's debut film?
10. In what country does the Three Colours trilogy take place?

ANSWERS

5. Slamdance
4. Robert Evans
3. Baltimore
2. The Draughtsman's Contract
1. Derek Jarman

10. France
9. sex, lies and videotape
8. Days of Heaven
7. New Jersey
6. TV Nation

1. What was director Todd Hayne's recreation of 1950s Hollywood melodramas called?

2. What was Edward Burns' directorial debut?

3. In which film does Kevin Spacey play an evil Hollywood producer?

4. Hilary Swank played Brandon Teena in which film?

5. Which Steve Buscemi film tells the story of an independent film shoot?

6. In Secrets and Lies, who plays the adopted daughter searching for her birth mother?

7. Neil LaBute's directorial debut was about two men conning a deaf girl – what was it called?

8. What Mike Leigh film takes place in real time during a party and ends in death?

9. Director John Sayles earns money to make his films by doing what?

10. Emily Watson's film debut was a Lars Von Trier film – what was it called?

1. Far From Heaven
2. Brothers McMullen
3. Swimming With Sharks
4. Boys Don't Cry
5. Living in Oblivion
6. Marianne Jean-Baptiste
7. In The Company of Men
8. Abigail's Party
9. Script-doctoring
10. Breaking the Waves

QUESTIONS

1. In what film does Paul Giamatti play a wine enthusiast facing a mid-life crisis?
2. What film studio was founded by the Weinstein Brothers?
3. Garden State was directed by and starred which TV comedy star?
4. Valmont was made a year after which film based on the same story?
5. In what Korean film does a man eat a live Octopus?
6. What Bollywood movie is an adaptation of Jane Austen's Pride and Prejudice?
7. For what film was Ellen Burstyn nominated for an Academy Award in 2000 for her performance as a weight loss pill-addicted mother?
8. Who directed The Passion of the Christ?
9. Mark Ruffalo and Laura Linney play estranged siblings in what 2000 drama?
10. What film did Billy Bob Thornton write, direct and star in?

ANSWERS

1. Sideways
2. Miramax
3. Zach Braff
4. Dangerous Liasions
5. Oldboy
6. Bride and Prejudice
7. Requiem for a Dream
8. Mel Gibson
9. You Can Count on Me
10. Sling Blade

QUESTIONS

1. What 1983 Louis Malle philosophical film takes place during real time at a dinner table?
2. What country does the film Amores Perros come from?
3. Before Sunset is the sequel to which film?
4. Reservoir Dogs was the debut film of which acclaimed director?
5. What is the name of the blind samurai featured in many Japanese films?
6. What was David Lynch's first film?
7. What film chronicles Edward Norton's last 24 hours before he goes to prison?
8. What actor played the title role in the Spike Lee film Malcolm X?
9. What was the sister film to the film Smoke?
10. Who plays the title role in Bad Lieutenant?

ANSWERS

1. My Dinner With Andre
2. Mexico
3. Before Sunrise
4. Quentin Tarantino
5. Zatoichi
6. Eraserhead
7. The 25th Hour
8. Denzel Washington
9. Blue in the Face
10. Harvey Keitel

QUESTIONS

1. Ralph Fiennes stars in a David Cronenberg film set in London's Kings Cross district, what is it called?
2. What Coen brothers film is a sepia-toned, Depression-era, musical version of Homer's Odyssey?
3. Who directed the films The Funeral and Bad Lieutenant?
4. What Spike Lee film is set around the Son of Sam killings?
5. Who plays the modern-day samurai in Ghost Dog: Way of the Samurai?
6. What is Billy Bob Thornton's profession in The Man Who Wasn't There?
7. In which film do Jude Law and Jennifer Jason Leigh enter a real-life video game?
8. What Academy-Award winner documentary tells the story of 'The Rumble in the Jungle' boxing match?
9. Who plays the title role in The King of New York?
10. Who plays Justine in The Good Girl?

ANSWERS

1. Spider
2. O Brother, Where Art Thou?
3. Abel Ferrara
4. Summer of Sam
5. Forest Whittaker
6. Barber
7. Existenz
8. When We Were Kings
9. Christopher Walken
10. Jennifer Anniston

1. Who directed Gosford Park?
2. Mulholland Drive stars which Australian actress?
3. What was Alexander Payne's film before Sideways?
4. What is the predominant colour in Amelie?
5. What drinks company does Bill Murray advertise in Lost in Translation?
6. To what does the title of the film 21 Grams refer?
7. What American issue does Bowling for Columbine tackle?
8. Bad Education is a film by which Spanish director?
9. What does Imelda Staunton do to 'help' girls in Vera Drake?
10. What Ken Loach film deals with an interracial romance between a Muslim man and a Catholic woman?

1. Robert Altman
2. Naomi Watts
3. About Schmidt
4. Green
5. Suntory
6. The supposed weight of the human soul
7. Guns
8. Pedro Almodovar
9. Abortions
10. A Fond Kiss

QUESTIONS

1. In what country does Y Tu Mama Tambien take place?
2. What was Julianne Moore's first collaboration with Far From Heaven director Todd Haynes?
3. Wes Anderson made a film loosely based on the life of Jacques Costeau – what is it called?
4. What was Zhang Yimou's follow up to Hero?
5. What country did Goodbye, Lenin! come from?
6. Who plays Billy Bob Thornton's boss in The Man Who Wasn't There?
7. What actress is Charlize Theron's lover in the film Monster?
8. What Mexican actress plays the famous painter Frida Khalo in the film Frida?
9. What French film tells the story of a marriage in reverse?
10. The Pianist takes place during what war?

ANSWERS

1. Mexico
2. Safe
3. The Life Aquatic with Steve Zissou
4. House of the Flying Daggers
5. Germany
6. James Gandolfini
7. Christina Ricci
8. Salma Hayek
9. 5 X 2
10. World War II

1. What does Morgan Spurlock eat every day for 30 days in Super Size Me?
2. Who plays Maggie Glyenhall's strict boss in the film Secretary?
3. In which Gus Van Sant film do Keanu Reeves and River Phoenix play male prostitutes?
4. Enduring Love is a film adaptation of a novel by which author?
5. In which Mike Leigh film does Timothy Spall play a London cabbie?
6. What Scottish actor directed The Magdalene Sisters?
7. What documentary tells the story of climbers Joe Simpson and Simon Yates and their Peruvian climb?
8. Which Lars Von Trier film stars Nicole Kidman and Lauren Bacall?
9. What actress stars in Sex & Lucia?
10. What film stars Michelle Williams and Anna Friel as old friends?

1. McDonalds
2. James Spader
3. My Own Private Idaho
4. Ian McEwan
5. All or Nothing
6. Peter Mullan
7. Touching the Void
8. Dogville
9. Paz Vega
10. Me Without You

ROUND 10
ALTERNATIVE CINEMA 10

QUESTIONS

1. Michael Moore endorsed a political candidate for President in the 2004 US election. Which one?
2. Amelie was set in which Paris neighbourhood?
3. What large film studio refused to release Fahrenheit 9/11 in 2004?
4. What is the highest-grossing independent film in history?
5. Which Russian film about a naval mutiny has a scene with a pram falling down steps and was referenced in The Untouchables?
6. Which black filmmaker rose to fame in 1989 with the controversial film Do the Right Thing?
7. Where is the largest film market in the world?
8. Musical dramas made in India are usually known by what name?
9. Y Tu Mama Tambien director Alfonso Cuaron directed what popular children's film in 2004?
10. What young Mexican director played Che Guevara in The Motorcycle Diaries?

ANSWERS

1. John Kerry
2. Montmartre
3. Disney
4. My Big Fat Greek Wedding
5. Battleship Potemkin
6. Spike Lee
7. India
8. Bollywood
9. Harry Potter and the Prisoner of Azkaban
10. Gael Garcia Bernal

QUESTIONS

1. Hero, House of Flying Daggers and Crouching Tiger Hidden Dragon have which actress in common?
2. What French film features two women on a travelling rampage of extreme sex and violence?
3. In which John Waters film does Kathleen Turner play a serial killer?
4. What French film was later remade in America as The Assassin starring Bridget Fonda?
5. What country does Babbette's Feast come from?
6. What Francois Truffaut film is based on a Ray Bradbury novel?
7. Who directed the film Wings of Desire?
8. What actor is most closely associated with director Werner Herzog?
9. What Italian film won the Oscar for Best Foreign film in 1988?
10. Which Italian director made Salo-120 Days of Sodom?

ANSWERS

10. Pier Paolo Pasolini
9. Cinema Paradiso
8. Klaus Kinski
7. Wim Wenders
6. Fahrenheit 451

5. Denmark
4. Le Femme Nikita
3. Serial Mom
2. Baise Moi
1. Ziyi Zhang

ROUND 10
ALTERNATIVE CINEMA 12

QUESTIONS

1. In which Japanese film do a group of school children fight each other to death?
2. Which two American directors financed Kurosawa's film Kagemusha?
3. In which Lars Von Trier film does Bjork star as a near-blind factory worker?
4. In The Seventh Seal, who does Max Von Sydow play chess with?
5. From which country is the film My Life as a Dog?
6. For which film was Imelda Staunton nominated for an Academy Award?
7. What Cuban film tells the story of the friendship between a gay man and a hardliner communist?
8. Who directed the film True Stories?
9. What was director Guillermo Del Toro's first film?
10. For what director did Antonio Banderas perform his first starring roles?

ANSWERS

1. Battle Royale
2. George Lucas and Francis Ford Coppola
3. Dancer in the Dark
4. Death
5. Sweden
6. Vera Drake
7. Strawberry and Chocolate
8. David Byrne
9. Cronos
10. Pedro Almodovar

1. What film tells the story of Anais Nin and Henry Miller?
2. Buena Vista Social Club chronicles a band from which country?
3. Maggie Glyenhall plays the masochistic employee of James Spader in what film?
4. Which American indie director made Limbo, Lone Star and Sunshine State?
5. What French film was remade as The Birdcage?
6. In which Almodovar film does Penelope Cruz play a nun with AIDS?
7. The Argentinean film Nine Queens was remade in Hollywood under which name?
8. Rabbit Proof Fence tells the story of two aborigine girls in what country?
9. During which sporting event does Belleville Rendezvous take place?
10. The Quiet American is based on a novel by which author?

1. Henry & June
2. Cuba
3. Secretary
4. John Sayles
5. La Cage Aux Folles
6. All About My Mother
7. Criminal
8. Australia
9. Tour de France
10. Graham Greene

ROUND 10
ALTERNATIVE CINEMA 14

1. Which Michael Winterbottom film chronicles the Manchester music scene?
2. Motorcycle Diaries tells the story of which revolutionary hero's early years?
3. Nicole Kidman starred in which Gus Van Sant film?
4. Lantana is a thriller from which country?
5. Which Ken Loach film follows a sixteen year old Scottish boy?
6. During which Presidential election campaign was Fahrenheit 9/11 released?
7. What does the protagonist of Sideways obsess over?
8. Sean Penn, Benicio del Toro and which actress star in 21 Grams?
9. Who stars as killer Aileen Wournos in the film Monster?
10. Who stars alongside Bill Murray in Lost In Translation?

10. Scarlett Johansson
9. Charlize Theron
8. Naomi Watts
7. Wine
6. 2004

5. Sweet Sixteen
4. Australia
3. To Die For
2. Che Guevara
1. 24 Hour Party People

QUESTIONS

1. What was Michael Moore's first film called?
2. What director made Stranger Than Paradise and Ghost Dog?
3. What are the biggest Independent film awards?
4. In which film does John Malkovich play himself?
5. Who wrote the films Eternal Sunshine of the Spotless Mind and Adaptation?
6. What famous film festival takes place in Park City, Utah?
7. Gods and Monsters tells the story of which Hollywood director?
8. Which Robert Rodriguez film was paid for by the director becoming a human guinea pig?
9. What was Happiness director Todd Solondz's first film?
10. Who starred in the film adaptation of Hunter S. Thompson's Fear and Loathing in Las Vegas?

ANSWERS

1. Roger & Me
2. Jim Jarmusch
3. Independent Spirit Awards
4. Being John Malkovich
5. Charlie Kaufman
6. Sundance
7. James Whale
8. El Mariachi
9. Welcome to the Dollhouse
10. Johnny Depp

ROUND 10
ALTERNATIVE CINEMA 16

QUESTIONS

1. In 2001 a film about a shy French girl conquered the world. What was her name?
2. Sweden's most famous filmmaker has been nominated for 9 Academy Awards – what is he called?
3. Name the director of Hong Kong action classics The Killer and Hard Boiled?
4. In the 1960s France experienced a film movement of young new filmmakers. What was this movement known as?
5. Marcello Mastroianni plays a playboy journalist in which 1961 Italian classic?
6. Chow Yun Fat and Michelle Yeoh star in this 2000 martial arts epic about the pursuit of a lost magic sword?
7. Insomnia was an Al Pacino film remade from a film by the same name from which country?
8. The Magnificent Seven is a remake of what Japanese film from 1954?
9. What Spanish filmmaker directed 'All About My Mother'?
10. What is the name of the German futuristic film by Fritz Lang from 1927?

ANSWERS

1. Amelie
2. Ingmar Bergman
3. John Woo
4. New Wave
5. La Dolce Vita
6. Crouching Tiger Hidden Dragon
7. Sweden
8. Seven Samurai
9. Pedro Almodovar
10. Metropolis

1. In what country does Lost In Translation take place?
2. What 1999 British film tells the story of the Pakistani Khan family in the 1970s?
3. What film did Gurinder Chadha direct before Bride and Prejudice?
4. Set in India during an arranged wedding, what is the name of Mira Nair's 2001 film?
5. 'All I need to make a film is a gun and a girl' – which famous French director said this?
6. In which Luis Bunuel film does Catherine Deneuve become a prostitute?
7. Which Hong Kong director made Chungking Express and In The Mood for Love?
8. Which Belgian film is about a film crew following a serial killer?
9. Which Spanish film was remade in Hollywood as Vanilla Sky?
10. What was the first of a series of films by Francois Truffaut featuring Jean Pierre Leaud as the character Antoine Doinel?

1. Japan
2. East is East
3. Bend it Like Beckham
4. Monsoon Wedding
5. Jean Luc-Godard
6. Belle de Jour
7. Wong Kar-Wai
8. Man Bites Dog
9. Open Your Eyes
10. The 400 Blows

QUESTIONS

1. Amores Perros and Y Tu Mama Tambien share which actor?
2. What is the sequel to Jean de Florette called?
3. What is Michael Moore's hometown, the setting of his film Roger and Me?
4. The Fog of War is a film about what former Secretary of Defence of the US?
5. Kurt & Courtney and Biggie & Tupac are documentaries by which British filmmaker?
6. In which film does Harvey Keitel attempt to 'deprogram' Kate Winslet?
7. What British filmmaker directed Secrets & Lies?
8. Le Haine is a French film set in which city?
9. Shallow Grave stars Kerry McFadden, Christopher Eccleston and which other actor?
10. What John Woo film features Chow Yun-Fat in a hospital shoot out?

ANSWERS

1. Gael Garcia Bernal
2. Manon des Sources
3. Flint, Michigan
4. Robert McNamara
5. Nick Broomfield
6. Holy Smoke
7. Mike Leigh
8. Paris
9. Ewan McGregor
10. Hard Boiled

ULTIMATE
BRAINBUSTER

SCIENCE

FICTION

ROUND 11
SCIENCE FICTION 1

1. In the X-Files, what is Scully and Mulder's boss called?
2. The War of the Worlds was a radio play and two films – who wrote the original story?
3. 'Danger, Will Robinson' is a classic line from which TV series?
4. Who plays Magneto in the X-Men films?
5. In Gattaca, who does Ethan Hawke borrow DNA from?
6. Where do the aliens land in Close Encounters of the Third Kind?
7. Frank Herbert's novel Dune was made into a film by which director?
8. Former model Milla Jovovich stars with Bruce Willis in which sci-fi action film?
9. What part of his body does Tom Cruise lose (and then replace) in Minority Report?
10. Who starred as Barbarella?

1. Skinner
2. H.G. Wells
3. Lost In Space
4. Ian McKellan
5. Jude Law
6. Devil's Tower
7. David Lynch
8. The Fifth Element
9. His eyes
10. Jane Fonda

202   FILM & TV QUIZ BOOK

ROUND 11
SCIENCE FICTION 2

1. In which decade was The Day The Earth Stood Still released?
2. In which film do seven people wake up inside an endless maze filled with traps?
3. Which Michael Crichton written film was directed by Robert Wise and features a deadly alien virus?
4. What British studio produced the Quartermass films?
5. Bruce Willis travels into the past to save the planet in which film?
6. What TV show, based on a film, stars former McGyver star Richard Dean Anderson?
7. How many Alien films have there been made?
8. In which film does Jodie Foster travel into space?
9. I, Robot is based on a book by which author?
10. Which science fiction film was the biggest box office hit of 1968?

1. 1950s (1951)
2. Cube
3. The Andromeda Strain
4. Hammer
5. Twelve Monkeys
6. Stargate SG-1
7. Five (Alien, Aliens, Alien 3, Alien Resurrection, Alien Vs. Predator)
8. Contact
9. Isaac Asimov
10. 2001: A Space Odyssey

QUESTIONS

1. Flash Gordon was a serial made into a film in 1980 with music by which rock band?
2. What famous special effects master worked on the films Clash of The Titans, One Million Years B.C. and Jason and the Argonauts?
3. Where did the aliens land in the film They Came From Beyond Space?
4. Which Japanese studio created the long-lasting series of Godzilla films?
5. What actor played Captain Nemo in 20,000 Leagues Under the Sea?
6. In which film does HG Wells pursue Jack the Ripper in modern day San Francisco?
7. In which film does Mick Jagger play a 21st Century Bounty Hunter chasing Emilio Estevez?
8. What was the computer in 2001: A Space Odyssey called?
9. Demon Seed was based on a novel by which author?
10. Hitchcock collaborator, Saul Bass, directed on science fiction film. What was it called?

ANSWERS

1. Queen
2. Ray Harryhausen
3. Cornwall
4. Toho
5. James Mason
6. Time After Time
7. Freejack
8. HAL 9000
9. Dean Koontz
10. Phase IV

ROUND 11
SCIENCE FICTION 4

QUESTIONS

1. Who voiced Darth Vader in Star Wars?
2. Who played the first Dr. Who?
3. The earliest science fiction film was a Jules Verne adaptation by which French director/magician?
4. In which century does the original Star Trek series take place?
5. Orson Welles came to fame through the radio broadcast of which H.G. Wells novel?
6. Star Trek and Star Trek The Next Generation. Name the three spin-offs from these series.
7. The Island of Lost Souls was remade in 1996 with what actor as Dr. Moreau?
8. What is Japan's most famous giant monster?
9. To What alien race does Spock belong?
10. What is Darth Vader's real name?

ANSWERS

1. James Earl Jones
2. William Hartnoll
3. George Melies
4. 23rd Century
5. War of the Worlds
6. Deep Space Nine, Voyager and Enterprise
7. Marlon Brando
8. Godzilla
9. Vulcan
10. Annakin Skywalker

1. George Clooney's Solaris was a remake of a 1972 Russian film by which director?
2. What was Jean Luc Godard's only science fiction film?
3. What 1995 Kevin Costner film was the most expensive ever made up to that point?
4. What creatures scream 'Exterminate! Exterminate!'?
5. What 1997 film features a race of killer bugs and a young army sent to another planet to eliminate them?
6. Who plays entrepreneur John Hammond in the film Jurassic Park?
7. What film's plot centres on a fake mission to Mars?
8. How many years after Alien 3 does Alien Resurrection take place?
9. In which Arnold Schwarzenegger film does he get cloned?
10. What cartoon series by the creator of The Simpsons is set in the 30th century?

1. Andrei Tarkovsky
2. Alphaville
3. Waterworld
4. The Daleks
5. Starship Troopers
6. Richard Attenborough
7. Capricorn One
8. 200
9. 6th Day
10. Futurama

ROUND 11
SCIENCE FICTION 6

QUESTIONS

1. Contact is based on the book by which author?
2. Who plays the Fifth Element in the film of the same name?
3. Who created The X Files?
4. What part of his face does Tom Cruise lose in Minority Report?
5. Men in Black ends in which New York neighbourhood?
6. What female rock star plays a large role in Mad Max Beyond Thunderdome?
7. Who stars in Equilibrium?
8. Where is Blade Runner set?
9. In what British TV series is 95% of humankind wiped away, leaving the remaining 5% to fend for themselves?
10. What monsters destroy most of the world in the film Reign of Fire?

ANSWERS

6. Tina Turner
7. Christian Bale
8. Los Angeles
9. Survivors
10. Dragons

1. Carl Sagan
2. Milla Jovovich
3. Chris Carter
4. His eyes
5. Queens

1. What is the name of the creator of the original Star Trek series?
2. What do these films have in common: Blade Runner, Total Recall, Minority Report?
3. What is Princess Leia's relation to Luke Skywalker in Star Wars?
4. What was the sequel to 2001: A Space Odyssey called?
5. What is the name of the actor who plays Doctor McCoy in the original Star Trek series?
6. How many Planet of The Apes films have been made (including remakes)?
7. What actor plays C3PO in the Star Wars films?
8. In which year does Blade Runner take place?
9. In Star Trek, who are the Enterprise's two fiercest enemies?
10. What was the first film to use computer generated effects?

1. Gene Roddenberry
2. Based on Philip K. Dick stories
3. Sister
4. 2010: The Year We Make Contact
5. Deforest Kelly
6. Six: Planet of the Apes, Beneath... Escape... Conquest... Battle for... and the 2001 remake
7. Anthony Daniels
8. 2019
9. Klingons and Romulans
10. Tron

ROUND 11
SCIENCE FICTION 8

QUESTIONS

1. What actor plays Morpheus in The Matrix films?
2. In which film is The White House blown to pieces by alien invaders?
3. What was George Lucas first film, now the name of his sound system?
4. What TV series created in 1999 features an astronaut thrown into the future in the middle of an intergalactic war?
5. What Ed Wood sci-fi film is generally considered the worst film ever made?
6. Fahrenheit 451 is based on a novel by which author?
7. How many seasons are there in TV sci-fi epic Babylon 5?
8. David Cronenberg's Existenz starred Jennifer Jason Leigh and which English actor?
9. Who played Blake in Blake's 7?
10. In which city does Robocop take place?

ANSWERS

1. Lawrence Fishburne
2. Independence Day
3. THX 1138
4. Farscape
5. Plan Nine From Outer Space
6. Ray Bradbury
7. Five
8. Jude Law
9. Gareth Thomas
10. Detroit

QUESTIONS

1. Where does the film Sphere take place?
2. What was the name of the animated stories released after The Matrix that expand on that film's story?
3. In which film does Jane Fonda perform a zero-gravity strip?
4. In which film set in 2019 does Arnold Schwarzenneger run for his life?
5. What classic sci-fi TV series did Rod Serling create in the 1950s?
6. What film was Disney's first foray into science fiction?
7. Solaris was the third collaboration between director Steven Soderberg and which actor?
8. What actor plays a scientist trying to stop Godzilla from destroying New York?
9. In what film do scientists go to the centre of the earth to save the planet?
10. Who played Luke Skywalker in the first three Star Wars films?

ANSWERS

1. Deep sea
2. Animatrix
3. Barbarella
4. Running Man
5. The Twilight Zone
6. The Black Hole
7. George Clooney
8. Matthew Broderick
9. The Core
10. Mark Hamill

ROUND 11
SCIENCE FICTION 10

1. Jeff Bridges is an alien in which John Carpenter film?
2. What actor stars alongside Will Smith in Men in Black?
3. For how many seasons did The X-Files run?
4. What is the name of Fritz Lang's futuristic 1927 German masterpiece?
5. What Star Trek style spoof starred Sigourney Weaver and Tim Allen?
6. In The Man Who Fell to Earth, which pop star plays an alien?
7. The Time Machine was remade in 2001 with which Australian actor?
8. What Sean Connery film is a remake of High Noon set in outer space?
9. What was the ultimate purpose of humans by the aliens in the TV series V?
10. In which city does Soylent Green take place?

1. Starman
2. Tommy Lee Jones
3. Nine
4. Metropolis
5. Galaxy Quest
6. David Bowie
7. Guy Pearce
8. Outland
9. Food
10. New York

1. How do the characters in The Matrix transport themselves from the real world into the matrix?
2. Where does the DNA used for cloning dinosaurs come from in Jurassic Park?
3. What model was the car used to time travel in the Back to The Future films?
4. What is the name of the spaceship in the first Alien film?
5 . What is the name of Doctor Who's time-travelling ship?
6. In Star Trek: Next Generation, which actor plays Captain Jean-Luc Picard?
7. What planet are The Daleks from?
8. Who was originally slated to play Han Solo?
9. Where did Special Agent Fox Mulder attend University?
10. Who played Doctor Who between 1974 and 1981?

1. Via telephone
2. From a mosquito trapped in Amber
3. Delorean DMC12
4. Nostromo
5. TARDIS (Time and Relative Dimensions in Space)
6. Patrick Stewart
7. Skaro
8. Christopher Walken
9. Oxford
10. Tom Baker

QUESTIONS

1. Who stars in the sci-fi film The Arrival?
2. In which film does astronaut Johnny Depp return slightly changed from outer space?
3. What British studio, better known for its horror titles, produced The Quartermass films?
4. What film is Invasion of the Body Snatchers set in high school?
5. In which sci-fi film is Charlton Heston one the few men still alive on Earth?
6. Philip K. Dick's story 'We Can Remember it For You Wholesale' became what Schwarzenegger film?
7. Who stars in the futuristic prison action film Fortress?
8. What was the final episode of The X-Files called?
9. In which Star Trek film do William Shatner and Patrick Stewart meet?
10. Who plays Elliot in the film E.T.?

ANSWERS

1. Charlie Sheen
2. The Astronaut's Wife
3. Hammer
4. The Faculty
5. The Omega Man
6. Total Recall
7. Christopher Lambert
8. The Truth
9. Star Trek: Generations
10. Henry Thomas

1. In Escape from New York, what has Manhattan been turned into?
2. Where does Will Smith find Tommy Lee Jones working in Men in Black 2?
3. What's the name of the second sequel to The Matrix?
4. What diminutive master Jedi trains Luke Skywalker?
5. What was the first dinosaur movie ever made?
6. In which film do a group of senior citizens recapture their youth through aliens?
7. In which era does the last film of the Back to the Future take place?
8. Who stars as The President of The United States in Deep Impact?
9. Who stars in The Butterfly Effect?
10. What film stars Bruce Dern trying to preserve the last surviving forest of Earth on a space ship?

1. A prison
2. Post Office
3. Matrix Revolutions
4. Yoda
5. The Lost World (1925)
6. Cocoon
7. Old West
8. Morgan Freeman
9. Ashton Kutcher
10. Silent Running

ROUND 11
SCIENCE FICTION 14

QUESTIONS

1. The Attack of the 50 ft Woman was remade with what actress as the giant?
2. What is the sequel to The Matrix called?
3. In which film does Kate Beckinsale play a vampire warrior?
4. Who plays Christopher Lambert's mentor in the film Highlander?
5. In what film does Dennis Quaid get trapped on a planet with an enemy alien?
6. In Cherry 2000, what is Melanie Griffith hired to track down?
7. What was Michael Winterbottom's foray into science fiction, starring Tim Robbins?
8. Who plays John Connor's love interest in Terminator 3: Rise of the Machines?
9. Sam Neill and which actor star in the sci-fi film Event Horizon
10. Who starred as Doctor Who in the feature films Doctor Who and the Daleks and Daleks: Invasion Earth 2150 AD?

ANSWERS

1. Daryl Hannah
2. Matrix Reloaded
3. Underworld
4. Sean Connery
5. Enemy Mine
6. A Robot Wife
7. Code 46
8. Claire Danes
9. Laurence Fishburne
10. Peter Cushing

1. What Stanley Kubrick project was later directed by Steven Spielberg?
2. In which film does Jennifer Lopez enter the mind of a serial killer?
3. Where does Keanu Reeves carry the data he transports in the film Johnny Mnenomic?
4. Jean Claude Van Damme is a time travelling police officer in which film?
5. Alien Nation stars James Caan and what actor as an alien police officer?
6. Who gets injected inside Martin Short in the film Innerspace?
7. Who stars as J. MacReady in The Thing?
8. In 2005, who took over the role of Dr. Who?
9. What TV producer created the series Buffy the Vampire Slayer, Angel and Firefly?
10. What sci-fi film marked both Drew Barrymore and William Hurt's film debuts?

10. Altered States
9. Joss Whedon
8. Christopher Eccleston
7. Kurt Russell
6. Dennis Quaid

5. Mandy Patinkin
4. Time cop
3. In his brain
2. The Cell
1. AI: Artificial Intelligence

ROUND 11
SCIENCE FICTION 16

QUESTIONS

1. Who stars as a roller ball hero in the futuristic film Roller Ball?
2. Death Race 2000 co-stars David Carradine and which action star?
3. What was the name of John Carpenter's first film, set in outer space?
4. In which TV show did Dr Spencer Quist say 'We must have more data'?
5. 'We have the technology – we can rebuild him' is from which series?
6. What organisation tries to prevent Alien invasion in the TV series UFO?
7. 'The Truth is Out there' is a slogan from which TV series?
8. Who plays the younger Obi Wan Kenobi in the Star Wars films?
9. What famous Hollywood director produced the alien abduction series Taken?
10. What do humans do to defeat the alien invasion in The War of the Worlds?

ANSWERS

1. James Caan
2. Sylvester Stallone
3. Dark Star
4. Doomwatch
5. Six Million Dollar Man
6. SHADO
7. The X Files
8. Ewan McGregor
9. Steven Spielberg
10. Human bacteria

1. In which film are people executed when they turn 30?
2. What is the name of the sequel to Pitch Black?
3. Who stars as The Duke of New York in the film Escape From New York?
4. Who sang a song for the film Silent Running?
5. Who stars as the computer programmer trapped inside a computer game in Tron?
6. What actor plays the title role in Hellboy?
7. In which film is a young orphan actually a top secret government experiment?
8. In Back to The Future, what is Michael J. Fox character's name?
9. In what year does Strange Days take place?
10. George Lucas created Industrial Light and Magic (ILM) for which film?

1. Logan's Run
2. Chronicles of Riddick
3. Isaac Hayes
4. Joan Baez
5. Jeff Bridges
6. Ron Perlman
7. DARYL
8. Marty McFly
9. 1999
10. Star Wars

ROUND 11
SCIENCE FICTION 18

QUESTIONS

1. For which film did George Lucas create the sound system THX?
2. Which Steven Spielberg science fiction film saved Columbia Studios from bankruptcy in 1977?
3. Which science fiction film was Natalie Wood's last?
4. What was the first film to feature full Computer Generated Images (CGI)?
5. Bruce Lee's son Brandon died on the set of which film?
6. What is the name of James Cameron's special effects company?
7. Which British TV sitcom was set in outer space?
8. What classic TV sci-fi series takes place in the in the third century of the second calendar?
9. What historic event took place the same day as the first transmission of a Dr. Who episode?
10. Which special effects wizard worked on 2001: A Space Odyssey, Close Encounters of the Third Kind and Star Trek: The Motion Picture?

ANSWERS

1. The Return of the Jedi
2. Close Encounters of the Third Kind
3. Brainstorm
4. The Young Sherlock Holmes
5. The Crow
6. Digital Domain
7. Red Dwarf
8. Blake's 7
9. Assassination of John F Kennedy
10. Douglas Trumbull

1. In what Samuel Fuller film does Richard Widmark get involved in a communist conspiracy?
2. In which year was The Night of the Living Dead first made?
3. Who plays the title role in the film The Crow?
4. What series of films follow a pair of dope-smoking friends?
5. What was the name of Quentin Tarantino's now-defunct DVD label for cult films?
6. What is the name of the family in Family Guy?
7. Who plays Feathers in Rio Bravo?
8. What actress plays Cici in Scream 2?
9. On which American TV network did Buffy the Vampire Slayer originate?
10. Which witch TV show starred Rose McGowan, Alyssa Milano and Shannon Doherty?

1. Pickup on South Street
2. 1968
3. Brandon Lee
4. Cheech & Chong
5. Rolling Thunder
6. Griffin
7. Angie Dickinson
8. Sarah Michelle Gellar
9. WB
10. Charmed

ROUND 12
CULT FILM & TELEVISION 2

1. In which Detective show did Pierce Brosnan star?
2. Jennifer Garner stars as Sydney Bristow in what TV series?
3. Keanu Reeves starred in two sci-fi comedies at the beginning of his career – what are they called?
4. Who is Alex's favourite composer in A Clockwork Orange?
5. Who plays Withnail in the film Withnail & I?
6. Who plays Donnie Darko's sister?
7. To what song does Michael Madsen torture the police officer in Reservoir Dogs?
8. What rock band produced and provided the music for Quadrophenia?
9. What is the name of the documentary about the filming of Apocalypse Now?
10. What 1936 classic anti-marijuana film is now a Broadway hit?

1. Remington Steele
2. Alias
3. Bill & Ted's Excellent Adventure and Bogus Trip
4. Beethoven
5. Richard E. Grant
6. Maggie Gyllenhaal
7. 'Stuck in the Middle With You' by Stealer's Wheal
8. The Who
9. Hearts of Darkness
10. Reefer Madness

1. What is the first (and second) rule of Fight Club?
2. Who plays Bateman in American Psycho?
3. What Bruce Willis film is based on a short film by documentary director Chris Marker?
4. In which film does Michael Douglas go on a killing rampage in Los Angeles?
5. Who plays former neo-Nazi Derek in American History X?
6. What member of the Rolling Stones starred in the 1970 film Performance?
7. What acclaimed British playwright wrote the script to the film Brazil?
8. What actor plays sensationalist reporter Wayne Gale in Natural Born Killers?
9. Vincent Vega is a character in which 1994 film?
10. What 1999 German film tells it story in a circle, over and over again?

1. You don't talk about Fight Club
2. Christian Bale
3. Twelve Monkeys
4. Falling Down
5. Edward Norton
6. Mick Jagger
7. Tom Stoppard
8. Robert Downey Jr
9. Pulp Fiction
10. Run Lola Run

1. What sci-fi TV show was created by Terry Nation and was first broadcast in 1978?
2. What TV show has Time Lords with two hearts?
3. In what year was The Hitchhikers Guide to the Galaxy first shown on TV?
4. What makes David Banner turn into The Hulk?
5. What actor starred in Baywatch and Knight Rider?
6. What was the name of th family in Lost in Space?
7. In which TV show does Robin Williams star as an alien?
8. In what decade was the TV version of The Planet of the Apes shown?
9. In which TV series does Scott Bakula travel time, arriving at a different era each episode?
10. Who played the Six Million Dollar Man?

1. Blake's 7
2. Dr Who
3. 1981
4. Anger
5. David Hasselhoff
6. Robinson
7. Mork & Mindy
8. 1970s (74)
9. Quantum Leap
10. Lee Majors

1. What sci-fi TV show did Gerry Anderson produce after UFO?
2. What actor played Mr. Spock on Star Trek?
3. In which Star Trek series is the captain a woman?
4. What TV producer created Lost in Space, Swiss Family Robinson, Time Tunnel and Voyage to the Bottom of the Sea?
5. What actress played Wonder Woman in the TV series?
6. What are X-Files fans known as?
7. Who played John Steed on The Avengers?
8. What show is the character Napoleon Solo from?
9. 'Your mission, Jim, should you choose to accept it…' is from which series?
10. Who plays the title role in Edward Scissorhands?

10. Johnny Depp
9. Mission: Impossible
8. The Man from U.N.C.L.E.
7. Patrick Macnee
6. X-Philes

5. Lynda Carter
4. Irwin Allen
3. Star Trek: Voyager
2. Leonard Nimoy
1. Space 1999

ROUND 12
CULT FILM & TELEVISION 6

1. What Australian actor played Jason King in Department S?
2. What is the name of the New York gang movie directed by Walter Hill in 1979?
3. What film features former TV-star Elizabeth Berkley as a stripper in Las Vegas?
4. Who played Emma Peel in the film version of The Avengers?
5. What 1996 film popularized the phrase 'You're so money'?
6. What TV show's main phrase was 'The Truth is Out There'?
7. 'We have the technology-we can rebuild him' – what series is this from?
8. What was the name of the car in Knight Rider?
9. What film features a giant signing plant?
10. What was the sequel to The Valley of the Dolls?

1. Peter Wyngarde
2. The Warriors
3. Showgirls
4. Uma Thurman
5. Swingers
6. The X Files
7. The Six Million Dollar Man
8. KITT
9. Little Shop of Horrors
10. Beyond the Valley of the Dolls

QUESTIONS

1. What film features the music and acting talents of The Ramones?
2. In which year is Blade Runner set?
3. Which actor starred in both Breakfast at Tiffany's and the A-Team?
4. Who played a gas-sucking maniac in Blue Velvet?
5. What actor played Batman in the 1960s TV series?
6. David Carradine played a Shaolin wandering the Old West in what series?
7. What cult director from Baltimore directed Pink Flamingos?
8. What city was Z-Cars set?
9. The Valley of the Dolls was based on a novel by what author?
10. Who is the star of Repo Man?

ANSWERS

1. Rock N' Roll High School
2. 2019
3. George Peppard
4. Dennis Hopper
5. Adam West
6. Kung Fu
7. John Waters
8. Liverpool
9. Jacqueline Susan
10. Emilio Estevez

ROUND 12
CULT FILM & TELEVISION 8

1. What director was famous for the large breasted women in his films?
2. Northern Exposure was set in what American state?
3. Who played Amanda Woodward in Melrose Place?
4. Which Kevin Bacon film features giant worms in the desert?
5. Who played Ed Wood in the Tim Burton film of the same name?
6. In what Ed Wood film do aliens invade Earth?
7. Where was Dynasty set?
8. What was the name of the car in The Dukes of Hazard?
9. What TV show features David Hasselhoff running in a swimming trunks in the opening credits?
10. What film features Tim Curry in full drag and Susan Sarandon as a naïve fiancee?

1. Russ Meyer
2. Alaska
3. Heather Locklear
4. Tremors
5. Johnny Depp
6. Plan Nine From Outer Space
7. Denver
8. General Lee
9. Baywatch
10. Rocky Horror Picture Show

QUESTIONS

1. Who was Prisoner Number 6?
2. What actress plays Emma Peel in The Avengers?
3. Who was the voice of Charlie in Charlie's Angels?
4. What TV show featured Bruce Lee as Kato, a limo chauffeur with martial arts skills?
5. What actor played Napoleon Solo in The Man from U.N.C.L.E.?
6. What actor stars in the Evil Dead series?
7. Francis Ford Coppola, Ron Howard and Martin Scorsese all started out working for what cult B-movies producer/director?
8. What New York-based studio created The Toxic Avenger and Surf Nazis Must Die?
9. What series featured Bruce Willis as a Private Investigator?
10. What country was Prisoner: Cell Block H from?

ANSWERS

1. Patrick McGoohan
2. Diane Rigg
3. John Forsythe
4. The Green Hornet
5. Robert Vaughn
6. Bruce Campbell
7. Roger Corman
8. Troma
9. Moonlighting
10. Australia

QUESTIONS

1. Who killed Laura Palmer in Twin Peaks?
2. What remake of The Wizard of Oz features Michael Jackson and Diana Ross?
3. What is the name of the film about Joan Crawford starring Faye Dunaway?
4. In They Live, what allows humans to detect aliens?
5. Who played Shaft in the TV version of the film?
6. In which action series did Roger Moore star as Simon Templar before his Bond era?
7. What sci-fi film did Duran Duran get their name from?
8. What is the name of the Australian apocalyptic film trilogy starring Mel Gibson?
9. 'I am not a number. I am a free man' – who said that?
10. What does IMF mean in Mission: Impossible?

ANSWERS

1. Her father – Leland Palmer
2. The Wiz
3. Mommie Dearest
4. Special sunglasses
5. Richard Roundtree
6. The Saint
7. Barbarella
8. Mad Max
9. Prisoner number six
10. Impossible Missions Force

QUESTIONS

1. Rob Reiner plays a documentary director following a British band on their disastrous US tour in which film?
2. What reggae singer starred in the film The Harder They Come?
3. What was the only film featuring punk band The Ramones?
4. Who wrote the songs featured in the film Harold & Maude?
5. What Ed Wood film featured an alien invasion that awakes the dead?
6. What human body part does Klye Mclachlan find in a field in Blue Velvet?
7. What was the first film shot in Odor-ama?
8. Stephen Dorff plays a psychotic filmmaker in John Waters' film about the film industry – what's its name?
9. Russ Meyer directed films with women that had what special characteristic?
10. In which New York neighbourhood does Requiem for a Dream take place?

ANSWERS

1. This is Spinal Tap
2. Jimmy Cliff
3. Rock N' Roll High School
4. Cat Stevens
5. Plan Nine From Outer Space
6. An ear
7. Polyester
8. Cecil B. Demented
9. Large breasts
10. Brooklyn

ROUND 12
CULT FILM & TELEVISION 12

QUESTIONS

1. What zombie film takes place mostly inside a shopping centre?
2. What 1977 bodybuilding documentary film stars Arnold Schwarzenegger?
3. What film movement did Melvin Van Peebles unleash with his film Sweet Sweetback's Badassssss Song?
4. Isaac Hayes won an Oscar for his theme song to which film?
5. In which film does Arnold Schwarzennegger play a sword-carrying brute?
6. What was The Monkees debut film?
7. What is Peter Fonda's character's name in Easy Rider?
8. What western stars Leonardo DiCaprio, Russell Crowe and Sharon Stone?
9. What French film features a giant wolf, a kung-fu fighting Native American, and Monica Belluci as a prostitute?
10. In which country were the original Ring films made?

ANSWERS

1. Dawn of the Dead
2. Pumping Iron
3. Blaxploitation
4. Shaft
5. Conan The Barbarian
6. Head
7. Captain America
8. The Quick and the Dead
9. Brotherhood of the Wolf
10. Japan

1. What famous action hero starred in Hercules in New York early in his career?
2. In an early scene in Pulp Fiction, what kind of burger does Samuel Jackson take a bite out of?
3. What flying monster is usually pitted against Godzilla?
4. What is the name of the strip club George Clooney and company arrive at in From Dusk Till Dawn?
5. Who plays Jimmy 'The Saint' in Things To Do In Denver When You're Dead?
6. In what city do the gangs fight it out on the streets in The Warriors?
7. What Japanese animation studio produced Spirited Away and Kiki's Delivery Service?
8. Who stars in the film One Million Years B.C.?
9. What TV actress stars in the American remake of The Grudge?
10. 'Stay away from the moors' is a warning from what film?

1. Arnold Schwarzenegger
2. Big Kahuna
3. Mothra
4. Titty Twister
5. Andy Garcia
6. New York City
7. Ghibli
8. Raquel Welch
9. Sarah Michelle Gellar
10. An American Werewolf in Paris

QUESTIONS

1. In the film Harley Davidson and the Marlboro Man Mickey Rourke plays the Harley Davidson man but who plays The Marlboro Man?
2. In what film does Jack Nicholson play a Hells Angels gang member?
3. Who starred in the motorcycle gang film The Wild One?
4. Who starred in and directed the film Driller Killer?
5. What Pink Floyd album is often played in conjunction with The Wizard of Oz?
6. In which film did Chloe Sevigny and Rosario Dawson get their start?
7. The Fall of the House of Usher is a film starring which actor?
8. Who starred in Beach Blanket Bingo?
9. What was Ed Wood's cross-dressing film called?
10. In what Sam Fuller film does a journalist enter a mental institution for research?

ANSWERS

1. Don Johnson
2. Hells Angels on Wheels
3. Marlon Brando
4. Abel Ferrara
5. The Dark Side of the Moon
6. Kids
7. Vincent Price
8. Frankie Avalon
9. Glen or Glenda
10. Shock Corridor

1. Faster Pussycat, Kill! Kill! Is a film by which director?
2. The Blues Brothers was a spin off from which TV comedy show?
3. What replaces Bruce Campbell's right arm in Army of Darkness?
4. Who wrote the original story on which A Clockwork Orange is based?
5. What German director remade Nosferatu in 1979?
6. Who played Sick Boy in Trainspotting?
7. Michelle Pfeiffer plays coke-addicted Elvira in which film?
8. New Jack City takes place in which New York neighbourhood?
9. Carlito's Way is Al Pacino's second turn as a Latin gangster – which was the first?
10. What was Johnny Depp's first starring role?

1. Russ Meyer
2. Saturday Night Live
3. Chainsaw
4. Anthony Burgess
5. Werner Herzog
6. Jonny Lee Miller
7. Scarface
8. Harlem
9. Scarface
10. Cry-Baby

ROUND 12
CULT FILM & TELEVISION 16

1. Who plays bail bondsman Max Cherry in Jackie Brown?
2. In which film does Edward Norton try to stop Project Mayhem?
3. In which David Lynch film does Isabella Rossellini play Dorothy?
4. 'Give yourself over to absolute pleasure' is a line from what musical?
5. Fear and Loathing in Las Vegas stars which actor as The Doctor?
6. Who plays Sailor in Wild at Heart?
7. What is Jeff Bridges known as in The Big Lebowski?
8. What Academy Award winning African-American actress was a regular guest star on Star Trek: The Next Generation?
9. Which actress starred in Foxy Brown?
10. What Jim Jarmusch film has a score by rock musician Neil Young?

1. Robert Forster
2. Fight Club
3. Blue Velvet
4. Rocky Horror Picture Show
5. Johnny Depp
6. Nicolas Cage
7. The Dude
8. Whoopi Goldberg
9. Pam Grier
10. Dead Man

ROUND 13

COMEDIES

1.   In which Marx Brothers film do they inadvertently start a war?
2.   Peter Sellers starred as which character in the Pink Panther movies?
3.   In which film does Dustin Hoffman pretend to be a woman?
4.   What is Mel Brooks' western spoof called?
5.   Dan Akroyd and which other actor were the Blues Brothers?
6.   Comedy director Harold Ramis starred as which character in Ghostbusters?
7.   Dr. Strangelove featured Peter Sellers in how many roles?
8.   Eddie Murphy moves from Africa to which city in Coming to America?
9.   In Bringing Up Baby, what feline species was Baby?
10. Kevin Kline won an Oscar for his role in which film?

1. Duck Soup
2. Inspector Closeau
3. Tootsie
4. Blazing Saddles
5. John Belushi
6. Egon
7. Two
8. New York City
9. Leopard
10. Fish Called Wanda

ROUND 13
COMEDIES 2

1. Who did Bob Hope star with in the Road To…films?
2. What film has Goldie Hawn joining the army?
3. What actor took Spencer Tracy's role in the remake of The Father of the Bride?
4. In which film do you find Sean Penn's Spicolli character?
5. What invention makes Steve Martin rich in The Jerk?
6. For which dark comedy film did Frances McDormand win the Oscar in 1996?
7. What was the first Gene Wilder - Richard Pryor pairing?
8. What were Laurel & Hardy's first names?
9. On what film did Tim Robbins and Susan Sarandon meet?
10. Who directed and starred in the original Nutty Professor?

1. Bing Crosby
2. Private Benjamin
3. Steve Martin
4. Fast Times at Ridgemont High
5. Nose glass holder
6. Fargo
7. Silver Streak
8. Stan & Oliver
9. Bull Durham
10. Jerry Lewis

QUESTIONS

1. Robin Williams is a DJ in what Vietnam-set film?
2. Who is Dudley Moore's leading lady in '10'
3. Who plays Mike Myers' love interest in Austin Powers: International Man of Mystery?
4. Jack Nicholson is an obsessive compulsive writer in which 1997 comedy?
5. Who plays Doc in the Back to the Future series?
6. Jennifer Garner wakes up 17 years older than she was in which comedy?
7. Who starred in Brewster's Millions?
8. William Hurt and Holly Hunter work in television in what film?
9. Catch-22 is based on a novel by which author?
10. What Kevin Smith film is set entirely inside a convenience store?

ANSWERS

1. Good Morning Vietnam
2. Bo Derek
3. Liz Hurley
4. As Good As It Gets
5. Christopher Lloyd
6. 13 Going on 30
7. Richard Pryor
8. Broadcast News
9. Joseph Heller
10. Clerks

ROUND 13
COMEDIES 4

1. In which slapstick film does Harold Lloyd hang from a clock on the side of a building?
2. David Brent is the main character of which TV comedy show?
3. Eric Idle, Michael Palin, Terry Gilliam, Terry Jones, Graham Chapman and what other actor formed Monty Python?
4. What action films spoof stars Arnold Schwarzenegger?
5. The team behind This is Spinal Tap are also behind what mockumentary about dog shows?
6. Father Ted is set on which Island?
7. Dylan Moran is a bookshop owner in what series?
8. Which silent comedy star made The General and Steamboat Bill, Jr.?
9. Which British studio was best known for its classic comedies?
10. Which Monty Python film deals with the King Arthur legend?

1. Safety Last
2. The Office
3. John Cleese
4. Last Action Hero
5. Best In Show
6. Craggy Island
7. Black Books
8. Buster Keaton
9. Ealing Studios
10. Monty Python and The Holy Grail

1. What American comedy TV show spawned the careers of Steve Martin, Bill Murray and Eddie Murphy?
2. What was Steve Martin's debut film?
3. In which show was did the Dead Parrot sketch appear?
4. In which film does Tom Cruise turn his parents' house into a whorehouse?
5. In which teen comedy do the protagonists vow to lose their virginity?
6. Who starred in the classic comedies It's A Gift and Bank Dick?
7. Films like Bringing Up Baby and It Happened One Night belong to what genre of comedy?
8. What is the name of the department store in Are You Being Served?
9. In what film does Woody Allen wake up in the future?
10. What actor played Blackadder?

10. Rowan Atkinson
9. Sleeper
8. Grace Brothers
7. Screwball
6. W.C. Fields

5. American Pie
4. Risky Business
3. Monty Python's Flying Circus
2. The Jerk
1. Saturday Night Live

ROUND 13
COMEDIES 6

1. You've Got Mail is a remake of what classic comedy?
2. In which café do Friends get together in?
3. What British sitcom takes place in outer space?
4. The comedy Best in Show takes place at what event?
5. What classic Hollywood slapstick studio forged the careers of Chaplin and Fatty Arbuckle?
6. Who was Jerry Lewis' comedic partner in seventeen films?
7. What was the first Carry On film?
8. What Italian comedian played the title role in Son of Inspector Closeau?
9. What Monty Python film deals with the story of Jesus Christ?
10. In which film is John Cusack a contract killer?

1. The Shop Around the Corner
2. Central Perk
3. Red Dwarf
4. Dog Show
5. Keystone
6. Dean Martin
7. Carry On Sergeant
8. Roberto Benigni
9. Monty Python and the Life of Brian
10. Grosse Pointe Blank

1. What film paired Dan Akroyd and Eddie Murphy?
2. Wayne's World originated as a sketch on what TV show?
3. Johnny Depp made a guest appearance on which British sketch show?
4. How many episodes of Fawlty Towers were made?
5. What was the name of Chris Morris' hugely controversial Newsnight-style spoof programme?
6. In which film does Jack Palance show Billy Crystal how to be a cowboy?
7. What was Robert De Niro's profession before retiring in Meet The Parents?
8. In which film does Ben Stiller play an incredibly vain supermodel?
9. Analyze This featured Robert DeNiro as a gangster in need of what?
10. Who does Julia Roberts try to steal her fiancé away from in My Best Friend's Wedding?

1. Trading Places
2. Saturday Night Live
3. The Fast Show
4. Twelve
5. Brass Eye
6. City Slickers
7. CIA Agent
8. Zoolander
9. Therapy
10. Cameron Diaz

ROUND 13
COMEDIES 8

1. In what town does The Office (UK) take place?
2. Mira Sorvino won an Oscar for which Woody Allen comedy?
3. In My Cousin Vinny, what is Marisa Tomei's area of expertise?
4. What film features Madonna playing baseball?
5. Who voices the baby in the Look Who's Talking films?
6. Hot Shots is mainly a spoof of what Tom Cruise film?
7. John Cleese plays Archie Leach in which 1987 comedy?
8. What was Shaun of the Dead star Simon Pegg's TV series called?
9. Who starred in the Naked Gun films?
10. Who plays Ferris Bueller?

10. Matthew Broderick
9. Leslie Nielsen
8. Spaced
7. A Fish Called Wanda
6. Top Gun

5. Bruce Willis
4. A League of Their Own
3. Cars
2. Mighty Aphrodite
1. Slough

QUESTIONS

1. Melanie Griffith played the part of which actress in the remake of Born Yesterday?
2. In which Chaplin film does his starved cabin mate imagine he is a chicken?
3. In what city is Ghostbusters set?
4. How high can the amp go in This Is Spinal Tap?
5. What is Holly Hunter's profession in Raising Arizona?
6. Who starred as The Thin Man in six films?
7. What actor is trapped in Groundhog Day?
8. In The Great Dictator, what does Hitler dance a ballet with?
9. Joel McRea meets which actress in Sullivan's Travels?
10. Cher won an Oscar for which 1987 comedy?

ANSWERS

1. Judy Holliday
2. Gold Rush
3. New York
4. 11
5. Police Officer
6. William Powell
7. Bill Murray
8. A world globe
9. Veronica Lake
10. Moonstruck

ROUND 13
COMEDIES 10

1. Who was Peter Cooke's famous collaborator?
2. What was the name of the League of Gentlemen's film?
3. Allo, Allo was set during which war?
4. What character on Friends used to be very obese?
5. Who does Tim end up with in the Christmas Special of The Office?
6. Who is Father Ted's faithful companion?
7. What series of films was Airplane! spoofing?
8. What was Monty Python's last film together?
9. Who starred in the comedies She Done Him Wrong and I'm No Angel?
10. In which English town is Alan Partridge based?

1. Dudley Moore
2. The League of Gentlemen's Apocalypse
3. World War II
4. Monica
5. Dawn
6. Father Dougal
7. Airport
8. Monty Python's The Meaning of Life
9. Mae West
10. Norwich

1. Who plays Sir Humphrey Appleby in Yes Minister?
2. In Keeping Up Appearances, how does Hyacinth Bucket pronounce her surname?
3. Who sang the theme tune to The Young Ones?
4. What was the name of the second Pink Panther movie?
5. 'You can't fight in here. This is the war room' is a line from which film?
6. Who wrote the novels on which the Bridget Jones films are based?
7. What American sitcom features a gay man and a straight woman living together?
8. In which film does Helen Mirren take her clothes off for charity?
9. What comedian created the TV series Phoenix Nights?
10. What links About a Boy and High Fidelity?

1. Nigel Hawthorne
2. Bouquet
3. Cliff Richards
4. A Shot in The Dark
5. Dr. Strangelove
6. Helen Fielding
7. Will & Grace
8. Calendar Girls
9. Peter Kay
10. Based on novels by Nick Hornby

ROUND 13
COMEDIES 12

1. What comedy features Ben Stiller as the owner of Globo Gym?
2. Which Victoria Wood comedy was a spoof of soap opera Crossroads?
3. In School of Rock, which actor turns a classroom into a rock band?
4. Who plays Hutch in the Starsky & Hutch film?
5. What is the name of the sequel to Shanghai Noon?
6. What TV comedy show stars Rowan Atkisnon as Inspector Fowler?
7. In Mean Girls what are the most popular girls in school known as?
8. What is the second sequel to American Pie called?
9. What classic comedy ends with the line 'Nobody's Perfect'?
10. In which film does Jim Carrey have God-like powers?

1. Dodgeball
2. Acorn Antiques
3. Jack Black
4. Owen Wilson
5. Shanghai Knights
6. Thin Blue Line
7. The Plastics
8. American Wedding
9. Some Like It Hot
10. Bruce Almighty

1. What are Jack Lemmon and Tony Curtis running from in Some Like it Hot?
2. What actress plays a soap star that falls for Dustin Hoffman in Tootsie?
3. What is the name of Stanley Kubrick's nuclear satire?
4. What is the name of the country Groucho Marx rules in Duck Soup?
5. What is Mel Brooks' western spoof called?
6. What actor plays Hawkeye in the film M.A.S.H.?
7. 'Are you seducing me, Mrs. Robinson' is a line from which film?
8. The Producers started as a film starring Gene Wilder and which other actor?
9. In what film does Groucho Marx stuff as many people as he can into a ship's stateroom?
10. What actor plays the Young Frankenstein?

10. Gene Wilder
9. A Night at the Opera
8. Zero Mostel
7. The Graduate
6. Donald Sutherland

5. Blazing Saddles
4. Freedonia
3. Dr. Strangelove
2. Jessica Lange
1. The mob

ROUND 13
COMEDIES 14

QUESTIONS

1. In which film does Cary Grant play stuffy zoologist David Huxley?
2. Jack Lemmon is one half of The Odd Couple – who is the other?
3. Which legendary slapstick star directed and acted in The General?
4. Cary Grant plays newspaper editor Walter Burns in which screwball classic?
5. In what film does Jack Lemmon lend out his flat for his boss's affairs?
6. Who plays Wanda in A Fish Called Wanda?
7. Judy Holliday is represented by lawyer Katherine Hepburn in what classic comedy?
8. Rob Reiner directed the Nora Ephron script starring Billy Crystal – what is it called?
9. What British comedian plays Tucker in There's Something About Mary?
10. In which film does Peter Sellers play a gardener addicted to television?

ANSWERS

1. Bringing Up Baby
2. Walter Matthau
3. Buster Keaton
4. His Girl Friday
5. The Apartment
6. Jamie Lee Curtis
7. Adam's Rib
8. When Harry Met Sally
9. Lee Evans
10. Being There

QUESTIONS

1. The Ghostbusters set up their headquarters in what type of building?
2. What is the name of the film that chronicles the band Spinal Tap on tour?
3. In Absolutely Fabulous, which is Edina Monsoon's profession?
4. Who plays the title role in the TV series Ally McBeal?
5. Who played Lonnie Goodbar in the series Porridge?
6. What actor played the 'Fonzie' on Happy Days?
7. In which American state did the Golden Girls live?
8. In which TV series does Dawn French play a vicar?
9. Who played Baldrick in the series Blackadder?
10. Rik Mayall was a writer and actor in which influential British TV series?

ANSWERS

1. An abandoned fire station
2. This is Spinal Tap
3. Public Relations Agent
4. Calista Flockhart
5. Richard Beckinsale
6. Henry Winkler
7. Florida
8. The Vicar of Dibley
9. Tony Robinson
10. The Young Ones

ROUND 13
COMEDIES 16

1. What do Nicolas Cage and Holly Hunter steal in Raising Arizona?
2. What was William Powell's dog's name in The Thin Man films?
3. What was Chaplin's first sound film?
4. In which American college-set film does John Belushi play Bluto?
5. William Powell and Carole Lombard star in which classic 1930s comedy?
6. In which comedy does a theatre company hide an American pilot in occupied Poland during WWII?
7. What actress plays the title role in the Western comedy Cat Ballou?
8. Marilyn Monroe is Tom Ewell's tempting neighbour in which film?
9. In which teenage comedy did Harrison Ford have a small part?
10. What was Greta Garbo's only comedy?

ANSWERS

5. My Man Godfrey
4. National Lampoon's Animal House
3. Modern Times
2. Asta
1. A baby

10. Ninotchka
9. American Graffiti
8. Seven Year Itch
7. Jane Fonda
6. To Be Or Not To Be

ROUND 13
COMEDIES 17

1. John Gielguld won an Oscar for his performance in which comedy?
2. In which film did Universal crossover their biggest horror stars with their biggest comedy stars for the first time?
3. In which Pink Panther film does Inspector Closeau go under cover at a nudist colony?
4. Who played Ednora in Bewitched?
5. In which city is Cheers set?
6. What fictional South Coast town was Dad's Army set in?
7. Who starred as the Fresh Prince of Bel Air?
8. What was Lucille Ball's husband's name?
9. The Mary Tyler Moore Show was spin off from which TV show?
10. What does MASH stand for?

1. Arthur
2. Abbot and Costello Meet Frankenstein
3. A Shot in The Dark
4. Agnes Moorehead
5. Boston
6. Walmington-on-Sea
7. Will Smith
8. Desi Arnaz
9. The Dick Van Dyke show
10. Mobile Army Surgical Hospital

ROUND 13
COMEDIES 18

1. 'I'm a Believer' was a song by which TV band?
2. In which decade was the series On the Buses shown?
3. What actor plays Roseanne's husband?
4. In which city is the film Diner set?
5. In which film does Betty Hutton wake up married to someone she doesn't know- and pregnant?
6. What famous vaudeville artist starred in The Court Jester and The Secret Life of Walter Mitty?
7. In which film is Kevin Costner a washed up baseball player training up and coming star Tim Robbins?
8. What was Gene Wilder and Richard Pryor's first film together?
9. What was Steve Martin's first film?
10. In which film does Julie Andrews disguise herself as a man?

1. The Monkees
2. 1970s (70-75)
3. John Goodman
4. Baltimore
5. The Miracle of Morgan Creek
6. Danny Kaye
7. Bull Durham
8. Silver Streak
9. The Jerk
10. Victor/Victoria

ROUND 14

DIRECTORS

1. What was the name of Orson Welles' theatre company?
2. What was Peter Bogdanovich's first film?
3. What form of transportation would Stanley Kubrick never travel in?
4. Alfred Hitchcock created which TV series?
5. For which computer company did Ridley Scott make a commercial in 1984?
6. Who directed the film Ali?
7. What Vietnam Veteran later became a multiple Academy Award winning writer-director?
8. What was Frank Sinatra's only attempt at directing?
9. For which Woody Allen film did Michael Caine win an Oscar?
10. What director made Rabbit Proof Fence and The Quiet American in 2003?

1. Mercury Theatre
2. Targets
3. Airplanes
4. Alfred Hitchcock Presents
5. Apple
6. Michael Mann
7. Oliver Stone
8. None But The Brave
9. Hannah and Her Sisters
10. Phillip Noyce

ROUND 14
DIRECTORS 2

QUESTIONS

1. Who directed both versions of The Ten Commandments?
2. Who is the Man with No Name?
3. Who directed (after many replacements) The Wizard of Oz?
4. Who directed Marlon Brando in Streetcar Named Desire and On The Waterfront?
5. What country did Stanley Kubrick move to in 1960?
6. George Lucas and Steven Spielberg worked together on which adventure trilogy?
7. What was David Lynch's first film?
8. Where are most of Sidney Lumet's films set?
9. What director married Judy Garland after directing her in Meet Me in St Louis?
10. Mike Nichols directed Elizabeth Taylor and Richard Burton in which film?

ANSWERS

10. Who's Afraid of Virginia Woolf?
9. Vincent Minnelli
8. New York City
7. Eraserhead
6. Indiana Jones

5. England
4. Elia Kazan
3. Victor Fleming
2. Clint Eastwood
1. Cecil B. DeMille

1. What 1915 film was directed by D.W. Griffith and was the most popular film ever until Gone with the Wind was released?
2. What is the first name of the director son of the great painter Renoir?
3. Whom did Woody Allen divorce in 1994?
4. What Robert Altman film set in the Korea war was turned into a successful TV series?
5. What 1968 film was remade by Tim Burton in 2001?
6. James Cameron directed the sequel to which sci-fi horror film in 1986?
7. John Cassavettes was nominated for a Best Actor Oscar for which film?
8. What was the name of the popular character created by Charlie Chaplin?
9. What are the first names of the Coen brothers?
10. What prolific Hungarian-born director helmed Casablanca?

1. Birth of a Nation
2. Jean
3. Mia Farrow
4. M.A.S.H.
5. Planet of the Apes
6. Alien
7. Dirty Dozen
8. The Tramp
9. Joel & Ethan
10. Michael Curtiz

ROUND 14
DIRECTORS 4

QUESTIONS

1. What maverick filmmaker directed The Wild Bunch?
2. Arthur Penn directed which groundbreaking film featuring Warren Beatty and Faye Dunaway?
3. Who directed Rebel Without a Cause?
4. What actress won an Oscar for Ken Russell's Women in Love?
5. What director turned Clint Eastwood into Dirty Harry?
6. What occupation did Quentin Tarantino famously engage in before being a director?
7. What medium did Orson Welles conquer before turning his sights on Hollywood?
8. Some Like it Hot was directed by which German-born director?
9. What British actor-turned-director made the Oscar-winning Gandhi?
10. What director was given the highest budget ever for a female director?

ANSWERS

1. Sam Peckinpah
2. Bonnie & Clyde
3. Nicholas Ray
4. Glenda Jackson
5. Don Siegel
6. Video store clerk
7. Radio
8. Billy Wilder
9. Richard Attenbourough
10. Kathryn Bigelow

DIRECTORS 5

1. What director made The Last Picture Show and Paper Moon?
2. What was Jack Nicholson's last film as a director?
3. Where do most of Woody Allen's films take place?
4. What Robert Altman film is centred on Hollywood?
5. Jimmy Stewart starred in how many Frank Capra films?
6. Who directed Dune?
7. What Australian director was responsible for the films Picnic at Hanging Rock and Gallipoli?
8. Marlon Brando directed one film- what was it called?
9. What pair of brothers is best known for directing gross-out comedies?
10. The Happy Days character Richie Cunningham was played by which director?

1. Peter Bogdanovich
2. The Two Jakes
3. New York
4. The Player
5. Two – Mr. Smith Goes to Washington and It's a Wonderful Life
6. David Lynch
7. Peter Weir
8. One Eyed Jacks
9. Farrelly Brothers
10. Ron Howard

QUESTIONS

1. What Sam Peckinpah film was banned in the UK until 2002?
2. What tale of two lovers on a killing spree did Arthur Penn direct?
3. What was John Huston and Humphrey Bogart's first film together?
4. What British director made The Red Shoes?
5. What famous director plays Victor Ziegler in Eyes Wide Shut?
6. What was the name of Martin Scorsese's film about Jesus Christ?
7. What director was Clint Eastwood's American mentor?
8. Who created the trilogy of The Man with No Name starring Clint Eastwood?
9. What film composer regularly scores all of Spielberg's films?
10. Oliver Stone directed Natural Born Killers, with a script by which young director?

ANSWERS

1. Straw Dogs
2. Bonnie & Clyde
3. Maltese Falcon
4. Michael Powell
5. Sidney Pollack
6. The Last Temptation of Christ
7. Don Siegel
8. Sergio Leone
9. John Williams
10. Quentin Tarantino

1. Who was credited with directing The Wizard of Oz: King Vidor or Victor Fleming?
2. What director made seven films with Marlene Dietrich?
3. What 30s and 40s action and westerns director wore an eye patch?
4. What Franz Kafka book did Orson Welles turn into a film?
5. What director made The Sound of Music and West Side Story?
6. How many Oscars did William Wyler win?
7. The actress Asia Argento is the daughter of which Italian horror director?
8. What was Richard Attenborough's first film as a director?
9. Where is Bernardo Betrolucci from?
10. Peter Bogdanovich filmed a sequel to The Last Picture Show in 1990 – what was it called?

1. Victor Fleming
2. Josef Van Sternberg
3. Raoul Walsh
4. The Trial
5. Robert Wise
6. Three
7. Dario Argento
8. Oh, What a Lovely War!
9. Italy
10. Texasville

ROUND 14
DIRECTORS 8

1. What comedy director/actor made Blazing Saddles and Young Frankenstein?
2. Where is director Jane Campion from?
3. Who composes most of the music in John Carpenter's films?
4. What Michael Cimino-directed film ruined United Artists in 1980?
5. How many months did the shoot for Apocalypse Now last?
6. What successful children's films director is named after a famous explorer?
7. What director created both the Nightmare on Elm Street and Scream franchises?
8. What profession did director Cameron Crowe pursue before filmmaking?
9. The Silence of the Lambs and Philadelphia were directed by whom?
10. Stanley Donen was a director best known for his work in what film genre?

10. Musicals
9. Jonathan Demme
8. Rock journalism
7. Wes Craven
6. Christopher Columbus

5. Sixteen
4. Heaven's Gate
3. John Carpenter
2. New Zealand
1. Mel Brooks

1.   Who directed Strangers on a Train?
2.   What was Martin Scorsese's first film with Robert DeNiro?
3.   What genre of films did John Ford mainly direct?
4.   What is the name of the director of Jackie Brown?
5.   John Carpenter rose to fame with which film about a serial killer and a babysitter?
6.   What science fiction film with Harrison Ford did Ridley Scott direct?
7.   What is the name of Ridley Scott's younger brother, also a director?
8.   What Rawhide actor later became a multiple Academy Award-winning director?
9.   What did the writer-director team of Michael Powell and Emeric Pressburger call themselves?
10.   What British director starred in a series of insurance adverts?

1. Alfred Hitchcock
2. Mean Streets
3. Westerns
4. Quentin Tarantino
5. Halloween
6. Blade Runner
7. Tony Scott
8. Clint Eastwood
9. The Archers
10. Michael Winner

ROUND 14
DIRECTORS 10

1. Sam Mendes won the Academy Award in 1999 for his debut film – what was it called?
2. What is the name of Francis Ford Coppola's daughter, also a director?
3. Michael Mann was nominated for Best Director in 2000 for a film about the tobacco industry – what was it called?
4. Who directed Top Gun?
5. Who directed both Bugsy Malone and Evita?
6. How many Best Director Academy Awards has Martin Scorsese won?
7. Fritz Lang, whom directed M and Metropolis, was from which country?
8. Battleship Potemkin was directed by which Russian director?
9. What Italian director is responsible for The Godfather trilogy?
10. What was the name of Oliver Stone's film based on his experiences in Vietnam?

1. American Beauty
2. Sofia
3. The Insider
4. Tony Scott
5. Alan Parker
6. None
7. Germany
8. Sergei Eisenstein
9. Francis Ford Coppola
10. Platoon

1. What is the name of Steven Spielberg's production company, named after his first short film?
2. What Akira Kurosawa film is the story of a rape told from four different points of view?
3. What was Francois Truffaut's first film?
4. Who is Spain's most famous director?
5. Kenneth Brannagh directed Mary Shelly's Frankenstein with which actor as the monster?
6. What director used to shoot guns on the sets of his films to keep everyone on edge?
7. How much did El Mariachi cost director Robert Rodriguez?
8. What Kevin Smith film caused controversy with the Church in America when released?
9. Martin Scorsese filmed the last concert of which rock n' roll band in 1976?
10. In which Steven Spielberg film does Francois Truffaut star as a French scientist?

1. Amblin Entertainment
2. Rashomon
3. The 400 Blows
4. Pedro Almodovar
5. Robert DeNiro
6. William Friedkin
7. $7000
8. Dogma
9. The Band
10. Close Encounters of the Third Kind

QUESTIONS

1. What director's name is most associated with the Pink Panther movies?
2. Peter Bogdanovich fell in love with which of his leading ladies in 1971?
3. What film did Oliver Stone make about the New York financial markets?
4. What was George Clooney's directorial debut?
5. What famous New York artist also directed many experimental films?
6. What German filmmaker worked for Hitler and made the film Triumph of the Will?
7. In which film did Tim Robbins direct Susan Sarandon and Sean Penn?
8. What Hong Kong-based filmmaker made the films In the Mood for Love and 2046?
9. What female director did James Cameron marry?
10. What music video director was married to Sofia Coppola?

ANSWERS

1. Blake Edwards
2. Cybil Shepherd
3. Wall Street
4. Confessions of a Dangerous Mind
5. Andy Warhol
6. Leni Riefenstahl
7. Dead Man Walking
8. Wong Kar Wai
9. Kathryn Bigelow
10. Spike Jonze

1. Which Indian-American director made The Sixth Sense?
2. Which music video filmmaker directed Eternal Sunshine of the Spotless Mind?
3. To which English actress is director Sam Mendes married?
4. Which American director sets many of his coming of age dramatic comedies in Baltimore?
5. To which actress was director Blake Edwards married?
6. What was Ben Stiller's directorial debut?
7. Director Rob Reiner is the son of what comedian?
8. For which film did Kevin Costner win the Best Picture Oscar in 1990?
9. What kind of westerns did Sergio Leone direct?
10. In what film did Laurence Olivier direct Marilyn Monroe?

1. M Night Shyamalan
2. Michel Gondry
3. Kate Winslet
4. Barry Levinson
5. Julie Andrews
6. Reality Bites
7. Carl Reiner
8. Dances With Wolves
9. Spaghetti
10. The Prince and the Showgirl

ROUND 14
DIRECTORS 14

QUESTIONS

1. What was the name of the TV show created by director David Lynch?
2. Who directed Barbra Streisand in The Way We Were?
3. What film, written by Quentin Tarantino, was directed by Tony Scott?
4. Orson Welles directed his wife in The Lady from Shanghai – what was her name?
5. Ian McKellan played which director in Gods and Monsters?
6. What New Jersey director created the characters Jay and Silent Bob?
7. Who was the youngest ever nominated director at the Academy Awards?
8. What actor/director founded the Sundance Film Festival?
9. What was Michael Moore's only fictional film?
10. Steven Spielberg grew up in which American state?

ANSWERS

5. James Whale
4. Rita Hayworth
3. True Romance
2. Sydney Pollack
1. Twin Peaks

10. Arizona
9. Canadian Bacon
8. Robert Redford
7. John Singleton
6. Kevin Smith

1. What film did Stanley Kubrick finish just before his death in 1999?
2. Who directed Nightmare on Elm Street and Scream?
3. Peter Jackson directed which adapted trilogy?
4. What young director made Boogie Nights and Magnolia?
5. Who directed Marlon Brando in Last Tango in Paris?
6. Jane Campion's The Piano was made in which country?
7. What writer/director wrote Taxi Driver?
8. What acclaimed documentary director made The Fog of War?
9. What was the follow up to About Schmidt by director Alexander Payne?
10. What famous video director made Being John Malkovich and Adaptation?

1. Eyes Wide Shut
2. Wes Craven
3. Lord of The Rings
4. Paul Thomas Anderson
5. Bernardo Bertolucci
6. New Zealand
7. Paul Schrader
8. Errol Morris
9. Sideways
10. Spike Jonze

ROUND 14
DIRECTORS 16

QUESTIONS

1. What was Hitchcock's first American film?
2. Who directed Citizen Kane?
3. What was Michael Cimino's follow up to The Deer Hunter?
4. For which film did Oliver Stone win his first Academy Award?
5. Roman Polanski fled the US in 1975 for which crime?
6. What was Guy Ritchie's film with his wife Madonna called?
7. What is the name of Ken Loach's film about a boy and his kestrel?
8. Mike Leigh was nominated for an Oscar in 1996 for which film?
9. Lawrence of Arabia was directed by whom?
10. Howard Hawks directed Humphrey Bogart and Lauren Bacall in their first film together – what was it called?

ANSWERS

1. Rebecca
2. Orson Welles
3. Heaven's Gate
4. Midnight Express
5. Statutory Rape
6. Swept Away
7. Kes
8. Secrets and Lies
9. David Lean
10. To Have and Have Not

1. What Robert Altman film takes place in the heart of the country music industry?
2. Who directed the first Batman film in 1989?
3. What classic Ealing comedy did the Coen brothers remake?
4. What was Francis Coppola's first film made at his studio Zoetrope?
5. What Hungarian-born director made Casablanca and Mildred Pierce?
6. Who directed Gone with the Wind and Wizard of Oz?
7. What director appeared in 37 of his films in cameo roles?
8. Who was John Huston's father, an Academy Award -winning actor?
9. What was Stanley Kubrick's profession before directing?
10. What black filmmaker made a film about Malcolm X?

1. Nashville
2. Tim Burton
3. The Ladykillers
4. One From The Heart
5. Michael Curtiz
§6. Victor Fleming
7. Alfred Hitchcock
8. Walter Huston
9. Photographer
10. Spike Lee

ROUND 14
DIRECTORS 18

QUESTIONS

1. Which Richard Attenborough film won Best Picture in 1982?
2. John Boorman directed an environmentally-conscious film about a boy lost in the Amazon – what's it called?
3. What British director, more famous for his Scotland-based films, directed 28 Days Later?
4. What is the name of the film Mike Leigh made about Gilbert & Sullivan?
5. In what Stephen Frears' film is John Cusack a record-store owner?
6. Who directed The Commitments?
7. Who directed The Devils, Women in Love and Tommy?
8. How Green Was My Valley was the first Best Picture Academy Award for which director?
9. What director discovered Lauren Bacall in a magazine?
10. John Cassavettes was an independent filmmaker, but also an actor – what World War II film did he star in?

ANSWERS

1. Gandhi
2. The Emerald Forest
3. Danny Boyle
4. Topsy Turvy
5. High Fidelity
6. Alan Parker
7. Ken Russell
8. John Ford
9. Howard Hawks
10. The Dirty Dozen

AWARD WINNERS

1. For which film did Gwyneth Paltrow win her first Best Actress Oscar?

2. After Gwyneth's weepy acceptance speech, which other Best Actress Oscar winner broke down in tears when winning for Monster's Ball?

3. What actress said 'you like me, you really like me' after winning her second Oscar in 1984?

4. What acclaimed director has lost out to both Kevin Costner and Robert Redford in the Best Director category?

5. As of 2005, what actress has been nominated 13 times for an Oscar?

6. What actress has won more Oscars than any other – as of 2005?

7. Tom Hanks got his first Oscar nomination for playing a child in which film?

8. What New York-based Academy Award-winning director has never been present at the ceremony to accept his statuettes?

9. Who did Marlon Brando send to accept his Oscar for him in protest in 1973?

10. In which year was the first Academy Award handed out?

1. Shakespeare in Love
2. Halle Berry
3. Sally Field
4. Martin Scorsese
5. Meryl Streep
6. Katherine Hepburn – 4 times
7. Big
8. Woody Allen
9. A Native American woman.
10. 1928

QUESTIONS

1. What organisation organises the Golden Globes Awards?
2. Michael Moore made a controversial speech when accepting an Oscar for which documentary in 2003?
3. What actor was nominated twice in the same year for two different films in 2004?
4. What director was nominated twice in the same year for two different films in 2000?
5. In which year were the first BAFTA's handed out?
6. Julia Roberts won an Oscar for her role in Erin Brokovich. In what year?
7. In which film was Nicole Kidman's Oscar winning performance in 2003?
8. Which Film Festival honours its top film with a Palm d'Or?
9. Mira Sorvino won an Oscar for her role in a film by which director?
10. What magazine throws the most famous Oscar after-party?

ANSWERS

1. Foreign Press Association in Hollywood
2. Bowling for Columbine
3. Jamie Foxx
4. Steven Soderbergh – for Traffic and Erin Brokovich
5. 1952
6. 2000
7. The Hours
8. Cannes Film Festival
9. Woody Allen
10. Vanity Fair

QUESTIONS

1. Marlon Brando and Robert De Niro won Oscars for their performance as which character?
2. What is American TV's most distinguished Award called?
3. Tom Hanks won an Oscar for Philadelphia – and then another the next year for which film?
4. What Award does the Academy give to outstanding producers, past winners have included Steven Spielberg and Clint Eastwood?
5. Al Pacino was nominated for Best Actor four times in a row – in which decade?
6. Which Sopranos actor won Best Actor Emmy in 2001?
7. As of 2005, only three black actors have won Best Actor Awards at the Oscars: Denzel Washington, Sidney Poitier and who else?
8. Joan Fontaine defeated her sister at the Oscars in 1941 – who is her sister?
9. Ryan O'Neal's daughter, Tatum, was nominated for which 1973 film in which she acted with her father?
10. What black comedian hosted the Academy Awards in 2005?

ANSWERS

1. Vito Corleone
2. Emmy
3. Forrest Gump
4. Irving Thalberg Award
5. 1970s (72-75)
6. James Gandolfini
7. Jamie Foxx
8. Olivia de Havilland
9. Paper Moon
10. Chris Rock

ROUND 15
AWARD WINNERS 4

QUESTIONS

1. What was the last black and white film before Schindler's List to win Best Picture?
2. As of 2005, how many Best Picture Awards has Steven Spielberg won?
3. Which Oscar-winning film had the largest number of extras?
4. What was the first film to win Best Picture at the Oscars?
5. As of 2005 who was the eldest best actress winner?
6. In which film did Jon Voight get his first nomination?
7. Which actor has been nominated consecutively for two posthumous Oscars?
8. Whose first award was for playing a character called Terry Malloy?
9. Which British actor won in the same year that Julie Andrews won for Mary Poppins?
10. Which British actor played Errol Flynn in The Aviator?

ANSWERS

1. The Apartment
2. Cabaret
3. One
4. Ghandi
5. Wings
6. Jessica Tandy
7. Midnight Cowboy
8. James Dean
9. Marlon Brando
10. Rex Harrison
11. Jude Law

1. As of 2005, what film has received the most Academy Awards for acting?
2. What is the name of Sean Penn's brother, also an actor?
3. For which film did Hilary Swank win her first Oscar?
4. Who is the British actress with most Oscar wins – as of 2005?
5. What actress did Tom Hanks marry?
6. What actress did Laurence Olivier marry after Vivien Leigh?
7. What was Robert DeNiro's directorial debut?
8. What film stars Academy Award winners Sean Penn and Kevin Spacey?
9. What playwright/actor did Jessica Lange marry?
10. For which film did Woody Allen win the Best Original Screenplay Oscar?

10. Hannah and her Sisters
9. Sam Shepard
8. Hurlyburly
7. A Bronx Tale
6. Joan Plowright

5. Rita Wilson
4. Vivien Leigh
3. Boys Don't Cry
2. Chris Penn
1. Network (Three)

QUESTIONS

1. Which writer/director has the most Oscar nominations for Original Screenplay?
2. Which screenwriter wrote Being John Malkovich, Adaptation and Eternal Sunshine of the Spotless Mind?
3. Which Oscar-winning writer published a tell-all book called Adventures in the Screen Trade?
4. Which Italian director is the most nominated foreign language filmmaker at the Oscars?
5. In which film did Robert DeNiro gain 60lbs to play boxer Jake LaMotta?
6. In 1982, E.T. lost Best Picture to which film?
7. Who was the only Asian actor to win a Best Supporting Actor Oscar?
8. What 1985 film starred then husband-and-wife Jack Nicholson and Anjelica Huston?
9. In which film does Kevin Costner build a baseball pitch?
10. Francis Ford Coppola and Oliver Stone both won Oscars as writers before they did as directors – for which films?

ANSWERS

1. Woody Allen
2. Charlie Kaufman
3. William Goldman
4. Federico Fellini
5. Raging Bull
6. Gandhi
7. Haing S. Ngor for The Killing Fields
8. Prizzi's Honor
9. Field of Dreams
10. Patton and Midnight Express

QUESTIONS

1. In which year did Katherine Hepburn and Barbra Streisand both win a Best Actress Oscar?
2. What was the first film to star Maggie Smith and Judi Dench together?
3. In which film does Sally Field play a union organiser?
4. Dianne Wiest won two Oscars for her performances in which director's films?
5. Oliver Stone won the Best Director Oscar twice in the 1980s – for which films?
6. What TV show is Academy Award winner James L. Brooks the executive producer of?
7. In which country was Milos Forman born?
8. Who was the first movie star-turned-director to win an Academy Award for Best Director?
9. Who was the first British Director to win an Academy Award?
10. In which city do the Academy Awards take place?

ANSWERS

1. 1968
2. Ladies in Lavender
3. Norma Rae
4. Woody Allen
5. Platoon and Born on the Forth of July
6. The Simpsons
7. Czechoslovakia
8. Robert Redford
9. David Lean
10. Los Angeles

ROUND 15
AWARD WINNERS 8

QUESTIONS

1. Since 1999, on which day do the Academy Awards take place?
2. Oscar host Billy Crystal played what in This is Spinal Tap?
3. Who was the youngest male recipient of a Best Actor Oscar in 2003?
4. In which film does Benicio Del Toro play an honest Mexican cop?
5. For which film did Michael Caine win an Academy Award in 1999?
6. In Good Will Hunting, what is Robin Williams' profession?
7. What painter does Daniel Day Lewis play in My Left Foot?
8. Which deaf actress won the Best Actress Oscar in 1986?
9. Emma Thompson won an Oscar for her performance in which film?
10. What country did Anthony Hopkins become a citizen of?

ANSWERS

1. Sunday
2. A mime-artist/caterer
3. Adrien Brody
4. Traffic
5. The Cider House Rules
6. Psychiatrist
7. Christy Brown
8. Marlee Matlin
9. Howard's End
10. United States

1. Which British Labour MP is a two-time Academy Award winner?
2. The Sundance Film Festival was founded by which Academy Award winning director?
3. The Tribeca Film Festival in New York was founded by which Academy Award winning actor?
4. What is the man on the Oscar statuette holding?
5. Before Chicago, which was the last musical to win Best Picture Academy Award?
6. What film beat Citizen Kane to Best Picture Oscar in 1941?
7. Which Hitchcock film was the only one to win a Best Picture Academy Award?
8. David O. Selznick produced which Academy Award winning 1939 film?
9. Lawrence Olivier won Best Picture Oscar in 1948 for which Shakespeare adaptation?
10. For which film did Walt Disney receive one large and seven small Oscar statuettes for?

1. Glenda Jackson
2. Robert Redford
3. Robert De Niro
4. A sword
5. Oliver!
6. How Green Was My Valley
7. Rebecca
8. Gone With the Wind
9. Hamlet
10. Snow White and the Seven Dwarfs

ROUND 15
AWARD WINNERS 10

1. In what year did Sidney Poitier win an Academy Award for Best Actor?
2. Who played the title role in the film Hud?
3. Which two actresses won Oscars for their roles in the film Who's Afraid of Virginia Woolf?
4. What musical won the Best Picture Award in 1968?
5. Which famous Award ceremony held a month before the Oscars has a Musical and Comedy category?
6. Goldie Hawn won an Oscar for her first film – what was it called?
7. In what year were The French Connection, The Last Picture Show and A Clockwork Orange all in competition at the Oscars?
8. How many Supporting actors were nominated for The Godfather II?
9. In 1978, two films about Vietnam were in competition at the Oscars – which ones?
10. Which British film won the Oscar in 1981 for Best Picture?

5. Golden Globes
4. Oliver!
3. Elizabeth Taylor and Sandy Dennis
2. Paul Newman
1. 1963

10. Chariots of Fire
9. Coming Home and Deer Hunter
8. Four
7. 1971
6. Cactus Flower

1. In what season of the year does the Cannes Film Festival take place?
2. Which Italian Film Festival awards Golden Lions?
3. Marcia Gay Harden won an Oscar in 200 for playing the wife of which famous American painter?
4. Jim Broadbent won an Oscar for playing the husband of which famous English writer?
5. Which two actresses were nominated as Best Supporting Actress for the film Chicago?
6. What is Martin Scorsese's biography of Howard Hughes called?
7. Gangs of New York starred which actor as Butcher Bill?
8. In which film did Nicole Kidman wear a prosthetic nose?
9. For which film did Tom Hanks grow a large beard?
10. In which city does Spain's most famous Film Festival take place?

1. Spring
2. Venice
3. Pollock
4. Iris Murdoch
5. Catherine Zeta Jones and Queen Latifah
6. The Aviator
7. Daniel Day Lewis
8. The Hours
9. Cast Away
10. San Sebastian

1. Which actress's award, for her role in Gone with the Wind, was sold for $510,000?
2. Which 1985 film had 11 nominations and not a single win?
3. Who has hosted the Academy Awards more than any other person?
4. Which 1987 film won all the categories for which it was nominated?
5. Marisa Tomei won an Oscar for which film?
6. What year were the Oscars shown on TV for the first time?
7. Which Oscar-winning film director was born Allen Stewart Konigsberg in 1935?
8. Better known as an Oscar-winning actor, who directed the film A Bronx Tale?
9. As of 2005, other than Tom Hanks, who is the only actor who has won back to back Best Actor Oscars?
10. What rock star won an Oscar for his song The Streets of Philadelphia?

10. Bruce Springsteen
9. Spencer Tracy
8. Robert DeNiro
7. Woody Allen
6. 1953

5. My Cousin Vinny
4. The Last Emperor
3. Bob Hope
2. The Colour Purple
1. Vivien Leigh

ROUND 15
AWARD WINNERS 13

1. Ben Hur and Titanic won 11 Oscars each – as of 2005, what is the only other film to win that many Oscars?
2. Who is the only actor to have received a posthumous Academy Award?
3. As of 2005, who is the most nominated actor in Oscar history?
4. Which French director did Jane Fonda marry in 1965?
5. Which Italian director did Ingrid Bergman live with after 1947?
6. Jack Nicholson, Meryl Streep and Frank Sinatra were born in which American state?
7. Which actor was Gene Hackman's roommate in the 1950s?
8. Which New York stage actress was nominated seven times before winning an Oscar on her eighth nomination in 1985?
9. James L. Brooks won three Oscars for writing, producing and directing which film?
10. As of 2005, which director has won more Oscars than any other?

1. Lord of The Rings: Return of the King
2. Peter Finch
3. Meryl Streep (13 nominations)
4. Roger Vadim
5. Roberto Rossellini
6. New Jersey
7. Dustin Hoffman
8. Geraldine Page
9. Terms of Endearment
10. John Ford – four times

QUESTIONS

1. 'Do I amuse you?' is a Joe Pesci line from which film?
2. Who directed and starred in the film Bugsy?
3. In which film does Al Pacino play a retired blind Colonel?
4. In Six Degrees of Separation, which actor does Will Smith pretend to be the son of?
5. How many Oscars did The Shawshank Redemption win?
6. Edward Norton was nominated for his first performance in which film?
7. What English film was nominated for Best Picture in 1994?
8. Who wrote the novel on which The Silence of the Lambs is based?
9. What Academy-Award winning composer wrote the music for Taxi Driver, Fahrenheit 451 and Psycho?
10. What Hollywood star produced One Flew over the Cuckoo's Nest?

ANSWERS

1. Goodfellas
2. Warren Beatty
3. Scent of a Woman
4. Sidney Poitier
5. None
6. Primal Fear
7. Four Weddings and a Funeral
8. Thomas Harris
9. Bernard Herrmann
10. Michael Douglas

1. Which famous producer was behind Chinatown, Godfather and Love Story?

2. What was the name of Steven Spielberg's epic black and white Holocaust film?

3. Two westerns won the Best Picture Oscar in the 1990s – which ones?

4. Which film about a World War II General won Best Picture Oscar in 1970?

5. Jeff Bridges and Ben Johnson were nominated for Best Supporting Oscars for which film?

6. Ryan O'Neal's daughter won an Oscar for which film?

7. In Kramer V. Kramer, which actress does Dustin Hoffman divorce?

8. What does BAFTA stand for?

9. 2001: A Space Odyssey was based on a story by which sci-fi author?

10. Laurence Olivier and Vivien Leigh were one of two married couples to both win Oscars as actors – who is the other couple?

1. Robert Evans
2. Schindler's List
3. Dances with Wolves and Unforgiven
4. Patton
5. Last Picture Show
6. Paper Moon
7. Meryl Streep
8. British Academy of Film and Television Arts
9. Arthur C. Clarke
10. Paul Newman and Joanne Woodward

ROUND 15
AWARD WINNERS 16

1. To which comedian is Oscar-winner Anne Bancroft married?
2. For which film did Donna Reed win an Oscar in 1953?
3. What was Kevin Kline's film debut?
4. Which Hospital TV drama did Denzel Washington star in for six years?
5. Before Adrien Brody, who was the youngest Best Actor winner?
6. Which famous tennis player did Tatum O'Neal marry in 1986?
7. Shirley MacLaine is the sister of which famous actor/director?
8. Which playwright wrote A Streetcar Named Desire?
9. Which Australian actress played Katherine Hepburn in The Aviator?
10. What was the only film of the 1970s to win Best Director but not Best Picture?

10. Cabaret
9. Cate Blanchett
8. Tennessee Williams
7. Warren Beatty
6. John McEnroe

5. Richard Dreyfuss
4. St. Elsewhere
3. Sophie's Choice
2. From Here To Eternity
1. Mel Brooks

ROUND 16

RANDOM

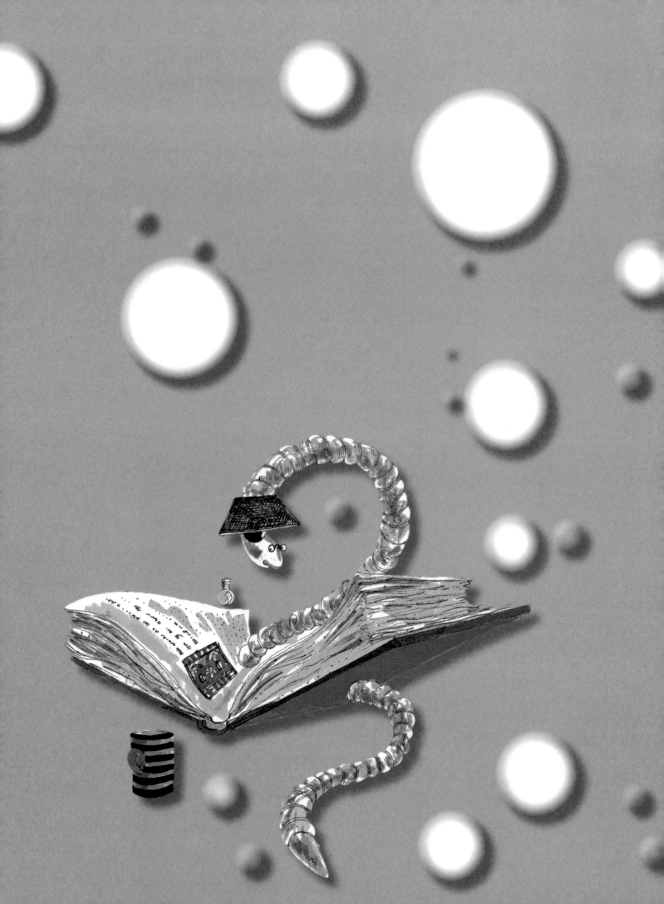

1. What 1986 Michael Mann film was remade as Red Dragon in 2002?
2. What actor was brought in to replace David Duchovny on the X Files?
3. What is the name of the film that tells the life story of Tina Turner?
4. Who starred alongside Susan Sarandon in Thelma and Louise?
5. Iceman and Maverick are characters in which 1986 film?
6. In which film does Tom Cruise's face become disfigured?
7.   Who directed Platoon?
8. Who starred as General Patton in the film Patton?
9. Where is Cary Grant originally from?
10. What was Clark Gable's last film?

1. Manhunter
2. Patrick Stewart
3. What's Love Got to do with it
4. Geena Davis
5. Top Gun
6. Vanilla Sky
7. Oliver Stone
8. George C. Scott
9. Bristol, England
10. The Misfits

ROUND 16
RANDOM 2

QUESTIONS

1. In which film did the character of Marion Crane famously die?
2. What is Mulder's first name in The X-Files?
3. In which 1988 film would you find a villain called Hans Gruber?
4. What film and television actress did Freddie Prinze Jr. marry in September 2002?
5. Archibald Leach is the real name of which actor?
6. Which actor was Madonna married to between 1985 and 1989?
7. Which 1999 film featured a sequence in which 7,900 rubber frogs were used?
8. In which country was Arnold Schwarzenegger born?
9. Who wrote, directed and starred in the 1996 film Sling Blade?
10. What was the name of film director George Lucas' dog which lent its name to a famous film character?

1. Psycho
2. Fox
3. Die Hard
4. Sarah Michelle Gellar
5. Cary Grant
6. Sean Penn
7. Magnolia
8. Austria
9. Billy Bob Thornton
10. Indiana

ANSWERS

QUESTIONS

1. What was the first name of John Travolta's character in Grease?
2. Which actor played John Malkovich in the film Being John Malkovich?
3. Who starred with Jack Lemmon in The Odd Couple?
4. Who played the title role in the 2002 film Spiderman?
5. What is the name of Tom Cruise's character in Mission Impossible?
6. Actor Brandon Lee died during the shooting of what film?
7. Which actress starred in Wild Things and the Scream trilogy?
8. In which country is most of the British horror film Dog Soldiers set?
9. Who played Frankie in the 1991 film Frankie and Johnny?
10. Who played Queen Elizabeth I in Blackadder II?

ANSWERS

1. Danny
2. John Malkovich
3. Walter Matthau
4. Tobey Maguire
5. Ethan Hunt
6. The Crow
7. Neve Campbell
8. Scotland
9. Michelle Pfeiffer
10. Miranda Richardson

QUESTIONS

1. What long-running British soap's episodes were shot in one take without any cuts?
2. Who did Bruce Lee lose the lead in TV series Kung Fu to?
3. Who plays James Garner's role in the film version of Maverick?
4. What Australian actor plays Crocodile Dundee?
5. What is the name of the sequel to 2001: A Space Odyssey?
6. What was the name of John Travolta's character in Saturday Night Fever?
7. Who played Hercules Poirot in the films 'Death on the Nile' and 'Appointment with Death'?
8. Who starred as Jim Hacker in 'Yes, Minister'?
9. What type of dog is Ren in 'The Ren and Stimpy Show'?
10. What four film stars formed United Artists in 1919?

ANSWERS

1. Crossroads
2. David Carradine
3. Mel Gibson
4. Paul Hogan
5. 2010: The Year We Make Contact
6. Tony Manero
7. Peter Ustinov
8. Paul Eddington
9. Chihuahua
10. Mary Pickford, Charlie Chaplin, Douglas Fairbanks Snr, D.W. Griffiths

QUESTIONS

1. In Thunderbirds, what colour was Lady Penelope's Rolls Royce?
2. What is Bob the Builder's girlfriend called?
3. In The Simpsons, which type of animal is Santa's Little Helper?
4. Which J. G. is the actress with the real name of Frances Gumm?
5. Who was Dr Richard Kimble accused of killing in the The Fugitive?
6. Who played Max Shreck in the 1992 film Batman Returns?
7. What was the surname of the professor played by Rex Harrison in the 1964 film My Fair Lady?
8. With which BBC TV show would you associate the character of David Brent?
9. In Tim Burton's film of The Planet Of The Apes who played the character which was played by Charlton Heston in the original?
10. Which famous film director directed the 1956 film The Killing?

ANSWERS

1. Pink
2. Wendy
3. Dog
4. Judy Garland
5. His wife
6. Christopher Walken
7. Higgins
8. The Office
9. Mark Wahlberg
10. Stanley Kubrick

ROUND 16
RANDOM 6

1.   Who directed the 1980 film The Elephant Man?
2.   In the movie world, how is Nicholas Kim Coppola better known?
3.   In July 2002, actress Angelina Jolie filed for divorce from which actor?
4.   Island Of Lost Dreams was part of the title to which 2002 movie sequel?
5.   Who directed the film Back to the Future?
6.   What was the sequel to the 1987 film Three Men and a Baby called?
7.   In which decade was the BBC TV series Grange Hill first broadcast - the 1970s or the 1980s?
8.   In which city is the 2001 film Mulholland Drive set?
9.   In which 1961 film did Jackie Gleason play Minnesota Fats?
10. In the film Austin Powers in Goldmember, what is the first name of Austin's father?

1. David Lynch
2. Nicolas Cage
3. Billy Bob Thornton
4. Spy Kids
5. Robert Zemeckis
6. Three Men and a Little Lady
7. 1970s
8. Los Angeles
9. The Hustler
10. Nigel

QUESTIONS

1. What nationality is comedian Steve Coogan's character Tony Ferrino?
2. What nationality was the actor Errol Flynn?
3. By what other name was the character of Oswald Cobblepot also known in the film Batman Returns?
4. In the film and TV series, what does the letter "A" stand for in M*A*S*H?
5. Who played Marc Anthony in the 1963 film "Cleopatra"?
6. Which real life gangster has been portrayed in films by Rod Steiger, Jason Robards and Robert De Niro?
7. What was the name of the wolf that befriended Kevin Costner in Dances with Wolves?
8. Which bird is the state bird of New Mexico, and also the name of a cartoon character?
9. Which actor famously said the line "Here's looking at you, kid" in the film Casablanca?
10. How many people does the Terminator played by Arnold Schwarzenegger kill in the film Terminator 2?

ANSWERS

1. Portuguese
2. Australian
3. The Penguin
4. Army
5. Richard Burton
6. Al Capone
7. Two Socks
8. Roadrunner
9. Humphrey Bogart
10. None

QUESTIONS

1. Which animal is displayed on the logo of MGM?
2. In what year did Eastenders first appear on television?
3. Which American rock star played a bus driver in the Spice Girls movie Spiceworld?
4. In 1991, Anthony Hopkins won the best actor Oscar for his performance in what film?
5. Which actress starred opposite Michael Douglas in the 1984 film Romancing The Stone?
6. In the cartoon series The Simpsons, what musical instrument does Lisa Simpson play?
7. What was the name of the murder victim at the beginning of the TV series Twin Peaks?
8. What was the title of the modern film version of Cyrano de Bergerac starring Steve Martin?
9. Who was the subject of the Mel Brooks film subtitled Men in Tights?
10. How many Oscars did the 1993 film Schindler's List receive?

ANSWERS

1. Lion
2. 1985
3. Meatloaf
4. Silence of the Lambs
5. Kathleen Turner
6. Saxophone
7. Laura Palmer
8. Roxanne
9. Robin Hood
10. Seven

1. Which of the Rocky films was released in 1982?
2. Which Oscar winning actor provided the voice of Hopper in the film A Bug's Life?
3. What sport features in the films He Got Game (1998), Flubber (1997) and Celtic Pride (1996)?
4. Who was author Ian Fleming's original choice to play James Bond on film, but who couldn't do it as he was already committed to a TV series?
5. Which sport is the 1988 film Bull Durham centred around?
6. Who directed the 1973 film American Graffiti?
7. In what film did Russell Crowe make his American debut, appearing with Sharon Stone, Leonardo DiCaprio and Gene Hackman?
8. In cinema, what came first - sound or colour?
9. In which film did Humphrey Bogart and Lauren Baccal first star together?
10. Which cartoon show included characters called Thelma and Shaggy?

1. Rocky III
2. Kevin Spacey
3. Basketball
4. Roger Moore
5. Baseball
6. George Lucas
7. The Quick and the Dead
8. Colour
9. To Have And Have Not
10. Scooby Doo

ROUND 16
RANDOM 10

QUESTIONS

1. What is the title of Steven Spielberg's first full length movie?
2. In which decade was the film The Exorcist released?
3. Who played studio executive Griffin Mill in the 1992 film The Player?
4. Which famous actor is the lead singer with the rock group 30 Odd Foot of Grunts?
5. Which comedy duo had first names Bud and Lou?
6. Who was the most popular cartoon character before Mickey Mouse?
7. Which actor links Legends of the Fall, Thelma & Louise and Ocean's Eleven?
8. In the film Who Framed Roger Rabbit, what is the name of Roger's wife?
9. In which film would you find the character of Dorothy Gale?
10. What type of monsters are at the centre of the films The Howling and Ginger Snaps?

ANSWERS

1. Duel
2. 1970's
3. Russell Crowe
4. Tim Robbins
5. Abbot & Costello
6. Felix the cat
7. Brad Pitt
8. Jessica
9. Wizard of Oz
10. Werewolves

ROUND 16
RANDOM 11

QUESTIONS

1. In what film did Basil Rathbone makes his first appearance as Sherlock Holmes?
2. In which 2000 film would you find the character of General Maximus Decimus Meridus?
3. In which country was Bob Hope born?
4. The film 10 Things I Hate About You is based on which Shakespeare play?
5. Which British actress has appeared in the films The Last Days Of Disco, Serendipity and Pearl Harbour?
6. What was the profession of the main character in the 2001 film The Man Who Wasn't There?
7. Which TV family live at 742 Evergreen Terrace?
8. Who directed the 1989 film Crimes And Misdemeanors?
9. What American film was released in France with a title which translated as Life, Love, and Cows?
10. The film All The President's Men is set around which political scandal?

ANSWERS

1. The Hound of the Baskervilles
2. Gladiator
3. England
4. The Taming of the Shrew
5. Kate Beckinsale
6. Barber
7. The Simpsons
8. Woody Allen
9. City Slickers
10. Watergate

ROUND 16
RANDOM 12

QUESTIONS

1. In which city was the soap opera Brookside set?
2. What is the name of the donkey in the film Shrek?
3. Who plays the title role in the 2002 film Mr Deeds?
4. Who played Fred in the 1994 film The Flintstones?
5. In which film does Harrison Ford play the character of Dr. Norman Spencer?
6. In the TV series Only Fools And Horses, what is the name of Boycie's wife?
7. What is Rachel's surname in Friends?
8. Which famous actor directed the film A River Runs Through It?
9. Who created The Muppets?
10. In which TV show would you find the character of Edina Monsoon?

ANSWERS

1. Liverpool
2. Donkey
3. Adam Sandler
4. John Goodman
5. What Lies Beneath
6. Marlene
7. Greene
8. Robert Redford
9. Jim Henson
10. Absolutely Fabulous

QUESTIONS

1. In which Australian city is the TV soap opera Neighbours set?
2. Robert De Niro played Max Cady in the Martin Scorsese film Cape Fear, but who played him in the 1962 version of the same film?
3. What was the name of Anthony Perkins' character in the film Psycho?
4. More widely recognised as an actor, who won the 1997 Oscar for best adapted screenplay?
5. Which Quentin Tarantino film was an adaptation of Elmore Leonard's novel called Rum Punch?
6. Which 1967 film starred Sidney Poitier, and had a theme tune performed by Lulu?
7. Which famous film director directed the second segment of the 1983 film Twighlight Zone: The Movie?
8. In which American city is the film Minority Report set?
9. What was the name of Basil's wife in Fawlty Towers?
10. The film Bicentennial Man starring Robin Williams was based on a story by which author?

ANSWERS

1. Erinsborough
2. Robert Mitchum
3. Norman Bates
4. Billy Bob Thorton
5. Jackie Brown
6. To Sir With Love
7. Steven Spielberg
8. Washington D.C.
9. Sybil
10. Isaac Asimov

QUESTIONS

1. Which actor in his autobiography said 'All I need to make a comedy is a park, a policeman and a pretty girl'?
2. In which American state is the 1953 film From Here to Eternity set?
3. In which country is the 1984 film The Killing Fields set?
4. In which film would you find Hatchet Harry and Barry the Baptist?
5. In which country was the BBC soap opera Eldorado set?
6. In which century was the 1995 film Judge Dredd set?
7. What won Best TV Comedy at the British Comedy Awards in 2002?
8. What is Truman's first name in the film The Truman Show?
9. What job did Helen Hunt's character do in the film As Good As It Gets?
10. In the film The Life of Brian, who played Brian?

ANSWERS

1. Charlie Chaplin
2. Hawaii
3. Cambodia
4. Lock Stock and Two Smoking Barrels
5. Spain
6. 22nd
7. The Office
8. Truman
9. Waitress
10. Graham Chapman

QUESTIONS

1. What was the name of the Blues Brothers star that died of a drug overdose in 1982?
2. Which famous cartoon character made his debut in the 1934 film The Wise Little Hen?
3. Who won a Golden Globe best actor award in 1999 for his performance in The Truman Show?
4. Who played the title role in the film Gandhi?
5. Which Bond villain had the first name of Auric?
6. Which film features the characters of Raymond and Charlie Babbitt?
7. In 1997, what took over from The Flintstones as the longest running prime-time animated series?
8. 'Eye of the Tiger' by Survivor was the theme song to which film?
9. The 1997 Spanish film Open Your Eyes was remade into which American film starring Tom Cruise?
10. In which film did Whoopi Goldberg win an Oscar playing the character of Oda May?

ANSWERS

1. John Belushi
2. Donald Duck
3. Jim Carrey
4. Ben Kingsley
5. Goldfinger
6. Rainman
7. The Simpsons
8. Rocky II
9. Vanilla Sky
10. Ghost

ROUND 16
RANDOM 15

QUESTIONS

1. In which European country was actor Bruce Willis born?
2. Which singer played Billie Holiday in the 1972 film Lady Sings The Blues?
3. Who played the title role in the Tim Burton film Ed Wood?
4. In which American city was the TV show Cheers set?
5. Which famous film actor once played the detective David Addison on television?
6. What two words usually begin the names of the episodes of the TV show Friends?
7. Who plays Dana Scully in The X-Files?
8. In the TV comedy series Frasier, what is Frasier's surname?
9. The film Saving Private Ryan is set around which war?
10. In which film would you find the Knights who say Ni?

ANSWERS

1. Germany
2. Diana Ross
3. Johnny Depp
4. Boston
5. Bruce Willis
6. The One
7. Gillian Anderson
8. Crane
9. World War 2
10. Monty Python's Holy Grail

1. Who plays Dr. Evil in the Austin Powers films?
2. Which character in the TV show Sesame Street lives in a trash can?
3. In the TV series Friends what is Joey and Chandler's favourite TV show?
4. Which film featured Robin Williams playing an English teacher called John Keating?
5. Who plays the Green Goblin in the 2002 film Spider-Man?
6. In which film did Clint Eastwood first appear as the Man with no name?
7. Who provided the voice of the title character in the 2001 film Shrek?
8. Who directed the 2001 film Harry Potter and the Philosopher's Stone?
9. In which city was actor Russell Crowe born?
10. In which film did Tom Hanks play a prison guard called Paul Edgecomb?

1. Mike Myers
2. Oscar
3. Baywatch
4. Dead Poets Society
5. Willem Dafoe
6. Fistful of Dollars
7. Mike Myers
8. Chris Columbus
9. Wellington, New Zealand
10. Green Mile

ROUND 16
RANDOM 18

1. What was the profession of Patrick Swayze's character in Ghost?
2. In which James Bond film does the character of Solitaire appear?
3. Who directed the films Faster, Pussycat! Kill! Kill! and Beyond The Valley Of The Dolls?
4. Which 1994 film featured Arnold Schwarzenegger playing a pregnant man?
5. What is the name of the motel in which Janet Leigh's character is killed in Psycho?
6. Which TV comedian plays the character of Alan Partridge?
7. Which historical figure did Mel Gibson play in the film Braveheart?
8. In the film Goldeneye, which British actor played agent 006?
9. What sport was the subject of the 1996 film Happy Gilmore starring Adam Sandler?
10. The film A.I. was completed by director Steven Spielberg, but which director started the project?

ANSWERS

1. Stockbroker
2. Live and Let Die
3. Russ Meyer
4. Junior
5. The Bates Motel
6. Steve Coogan
7. William Wallace
8. Sean Benn
9. Golf
10. Stanley Kubrick

1. Which two films that have won best picture Oscars in the 1990s have types of animals in their title?
2. Before finding fame as a film director, who played an Elvis impersonator in the last episode of The Golden Girls?
3. Who directed the film Almost Famous based on his own experiences as a rock reporter?
4. For which film did Henry Fonda receive his last Oscar?
5. In which country were the recent Lord of the Rings movies filmed?
6. In which English city is the film The Full Monty set?
7. Who directed the Oscar-winning film The English Patient?
8. Who directed the film Dr Strangelove?
9. Which TV show featured the character of DS Jane Penhaligon?
10. Jessica Tandy won the best Actress Oscar in 1989 for which film?

5. New Zealand
4. On Golden Pond
3. Cameron Crowe
2. Quentin Tarantino
1. Silence of the Lambs and Dances with Wolves

10. Driving Miss Daisy
9. Cracker
8. Stanley Kubrick
7. Anthony Minghella
6. Sheffield

ROUND 16
RANDOM 20

QUESTIONS

1. Which Monty Python star was given a lifetime achievement award at the British Comedy Awards in 2002?
2. Which member of the A Team was a pilot?
3. Which director's first film was called THX 1138?
4. In Doctor Who, what was the name of the chief of the Daleks?
5. Which actor played a character that was frozen for 50 years in the film Forever Young?
6. Whose last film as a director was Family Plot?
7. Which actor was one of the Dirty Dozen and also one of the Magnificent Seven?
8. Which actress won three BAFTAs in the 1990s for playing a policewoman on TV?
9. Better known in the world of music, who played the leading role in the film Buster?
10. Which real-life brothers starred opposite Michelle Pfeiffer in the 1989 film The Fabulous Baker Boys?

ANSWERS

1. Michael Palin
2. Murdoch
3. George Lucas
4. Davros
5. Mel Gibson
6. Hitchcock
7. Charles Bronson
8. Helen Mirren
9. Phil Collins
10. Jeff and Beau Bridges

ULTIMATE BRAINBUSTER

THE BRAINBUSTERS

Two pages of the hardest quiz questions on Earth.

ROUND 17
THE BRAINBUSTERS

QUESTIONS

1 Who was the first black actress to win an Academy Award?
2 'One Flew Over the Cuckoo's Nest' is based on a novel by which author?
3 What is Woody Allen's real name?
4 Which actor in David Byrne's 'True Stories' took his own life?
5 What blacklisted screenwriter did Kirk Douglas hire to write Spartacus?
6 In what area in Mexico was Titanic mostly filmed?
7 What Orson Welles-directed Booth Tarkington film adaptation was recut by the studio?
8 What western legend does My Darling Clementine tell the story of?
9 Who was Billy Wilder's frequent writing partner?
10 Dialogue from the film 'Performance' was sampled by which Mancunian band?

ANSWERS

1. Hattie McDaniel
2. Ken Kesey
3. Allen Konigsberg
4. Spalding Gray
5. Dalton Trumbo
6. Baja California
7. The Magnificent Ambersons
8. Wyatt Earp
9. I.A.L. Diamond
10. Happy Mondays

QUESTIONS

1 What was John and Walter Huston's only collaboration on film?

2 Who played Indiana Jones in the TV series?

3 What Jazz musician did Clint Eastwood make a film about?

4 Who screams 'Top of the world, Ma' in the film White Heat?

5 Which actress had a British Labour Party leader as a grandfather and a son involved with the Charles Manson Cult?

6 What film was inspired by the real-life killings by Charles Starkweather in the 1950s?

7 In what science fiction cult film does Robert DeNiro play Harry Tuttle?

8 What South American country do Butch Cassidy and the Sundance Kid flee to?

9 What war event frames From Here to Eternity?

10. What industrial building fascinates Ralph Fiennes' character in the Cronenberg film Spider?

ANSWERS

1. Treasure of Sierra Madre
2. Sean Patrick Flannery
3. Charlie Parker – Bird
4. James Cagney
5. Angela Lansbury
6. Badlands
7. Brazil
8. Bolivia
9. Pearl Harbor attack
10. A gasometer